Fashionable Clothing from the Sears Catalogs

early 1950s

Desire Smith

Schiffer Publishing Ltd

4880 Lower Valley Road, Atglen, PA 19310

Library of Congress Catalog Card Number: 97-81064

Book design by Blair Loughrey

Typeset in Futura BK BT 11/12.5 8/9
ISBN: 0-7643-0519-0
Printed in Hong Kong
1 2 3 4

Published by Schiffer Publishing Ltd.
4880 Lower Valley Road
Atglen, PA 19310
Phone: (610) 593-1777; Fax: (610) 593-2002
E-mail: Schifferbk@aol.com
Please write for a free catalog.
This book may be purchased from the publisher.
Please include $3.95 for shipping.
Try your bookstore first.

We are interested in hearing from authors
with book ideas on related subjects.

Contents

Preface

My parents moved from New York to rural Pennsylvania in the late 1940s. My mother quickly caught on to the custom of placing last season's Sears catalogs in the outhouse, which she called the "fancy house" and held out as one of the seven wonders of the world to her weekend guests from New York. My mother, not understanding the full folklore of this custom, for years placed every outdated catalog in the "fancy house," which, since we had indoor plumbing, was not frequented often enough to use the catalogs to their best advantage! As a little girl I spent a lot of time reading the outdated catalogs stacked there. Even today at estate auctions in rural areas such a stash of Sears' catalogs is likely to turn up, but I have to admit that my mother had the largest collection I've ever seen right there in the "fancy house."

Many items from the early 1950s Sears catalogs are now collectible: apparel, radios, bicycles, clocks, watches, fountain pens, cameras, furniture, lamps, bathroom scales, hampers, toys, games, cookware—the list goes on. The table of contents for each catalog is printed on its spine, a kind of index to the past!

Introduction

Fashionable Clothing from the Sears Catalogs, Early 1950s, focuses on wearable, collectible, vintage clothing and accessories. The Sears catalogs provide the basis for a comprehensive study of the fashions of this period. There is no guessing as to the dates of the clothing. We know the exact dates. The descriptions are unsurpassed in terms of detail and accuracy.

Virtually all clothing from the 1950s is collectible. Those items which epitomize the trends of the era are the most exciting to collectors. For example, felt circular skirts with poodle appliqués are always in demand. Equally interesting to collectors are the beautifully tailored gabardine suits, jackets, toppers, and swing coats of the early 1950s. The wonderfully styled handbags, especially variations on the box style, have become very popular in recent years. The market is strong for bathing suits, bathrobes, nightgowns, riding wear, dungarees, pedal pushers, shorts, blouses, dresses, gowns, hats, gloves, and shoes for women; smoking jackets, robes, boots and shoes, swimwear, shirts, jackets, suits, hats, ties, and even socks for men. There is a growing and very specialized market for sportswear, especially denim jeans and jackets, and rubber-soled canvas shoes.

Although vintage clothing from the early 1950s is available today, it is getting increasingly hard to find. Many closets and attics have been emptied, and the old clothing discarded or donated. Estate sales rarely offer large quantities of 1950s clothing, and the auction houses that hold specialized clothing sales offer only couture clothing of the period. Collectors who enjoy wearing the styles will continue looking for these vintage garments and accessories at quality vintage clothing shops and specialized shows. Those who love the flattering, wearable 1950s garments will not give up the hunt!

Fashionable Clothing from the Sears Catalogs, Early 1950s, showcases some of the most collectible and desirable garments and accessories of the period. Not only do the catalog descriptions help collectors understand the styles, they also give a complete analysis of the textiles used. This book focuses on fashions for women, but also includes examples of *collectible* fashions for men and children. Author's comments, designed to clarify and enhance, are interspersed with descriptive captions throughout the book, and a current value guide is placed next to the original Sears' prices.

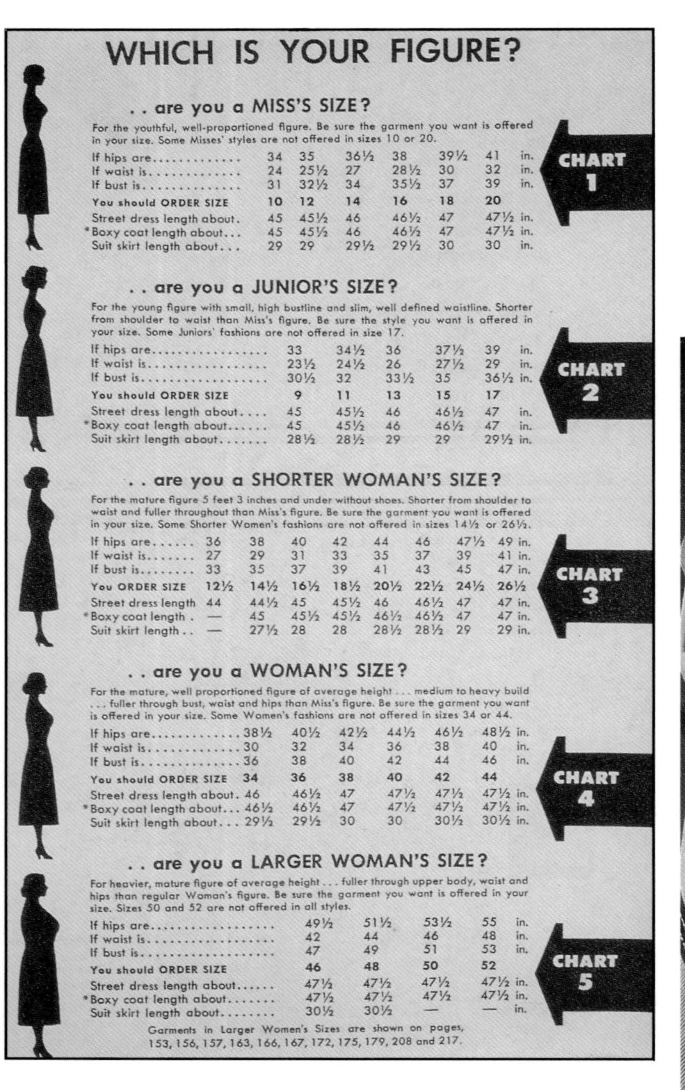

WHICH IS YOUR FIGURE?

. . are you a MISS'S SIZE?

For the youthful, well-proportioned figure. Be sure the garment you want is offered in your size. Some Misses' styles are not offered in sizes 10 or 20.

If hips are	34	35	36½	38	39½	41	in.
If waist is	24	25½	27	28½	30	32	in.
If bust is	31	32½	34	35½	37	39	in.
You should ORDER SIZE	**10**	**12**	**14**	**16**	**18**	**20**	
Street dress length about	45	45½	46	46½	47	47½	in.
*Boxy coat length about	45	45½	46	46½	47	47½	in.
Suit skirt length about	29	29	29½	29½	30	30	in.

CHART 1

. . are you a JUNIOR'S SIZE?

For the young figure with small, high bustline and slim, well defined waistline. Shorter from shoulder to waist than Miss's figure. Be sure the style you want is offered in your size. Some Juniors' fashions are not offered in size 17.

If hips are	33	34½	36	37½	39	in.
If waist is	23½	24½	26	27½	29	in.
If bust is	30½	32	33½	35	36½	in.
You should ORDER SIZE	**9**	**11**	**13**	**15**	**17**	
Street dress length about	45	45½	46	46½	47	in.
*Boxy coat length about	45	45½	46	46½	47	in.
Suit skirt length about	28½	28½	29	29	29½	in.

CHART 2

. . are you a SHORTER WOMAN'S SIZE?

For the mature figure 5 feet 3 inches and under without shoes. Shorter from shoulder to waist and fuller throughout than Miss's figure. Be sure the garment you want is offered in your size. Some Shorter Women's fashions are not offered in sizes 14½ or 26½.

If hips are	36	38	40	42	44	46	47½	49	in.
If waist is	27	29	31	33	35	37	39	41	in.
If bust is	33	35	37	39	41	43	45	47	in.
You ORDER SIZE	**12½**	**14½**	**16½**	**18½**	**20½**	**22½**	**24½**	**26½**	
Street dress length	44	44½	45	45½	46	46½	47	47	in.
*Boxy coat length	—	45	45½	45½	46½	46½	47	47	in.
Suit skirt length	—	27½	28	28	28½	28½	29	29	in.

CHART 3

. . are you a WOMAN'S SIZE?

For the mature, well proportioned figure of average height . . . medium to heavy build . . . fuller through bust, waist and hips than Miss's figure. Be sure the garment you want is offered in your size. Some Women's fashions are not offered in sizes 34 or 44.

If hips are	38½	40½	42½	44½	46½	48½	in.
If waist is	30	32	34	36	38	40	in.
If bust is	36	38	40	42	44	46	in.
You should ORDER SIZE	**34**	**36**	**38**	**40**	**42**	**44**	
Street dress length about	46	46½	47	47½	47½	47½	in.
*Boxy coat length about	46½	46½	47	47½	47½	47½	in.
Suit skirt length about	29½	29½	30	30	30½	30½	in.

CHART 4

. . are you a LARGER WOMAN'S SIZE?

For heavier, mature figure of average height . . . fuller through upper body, waist and hips than regular Woman's figure. Be sure the garment you want is offered in your size. Sizes 50 and 52 are not offered in all styles.

If hips are	49½	51½	53½	55	in.
If waist is	42	44	46	48	in.
If bust is	47	49	51	53	in.
You should ORDER SIZE	**46**	**48**	**50**	**52**	
Street dress length about	47½	47½	47½	47½	in.
*Boxy coat length about	47½	47½	47½	47½	in.
Suit skirt length about	30½	30½	—	—	in.

CHART 5

Garments in Larger Women's Sizes are shown on pages, 153, 156, 157, 163, 166, 167, 172, 175, 179, 208 and 217.

Note: Many collectors are puzzled about how sizing worked in the 1950s. This "Which Is Your Figure?" chart from the Spring/Summer 1953 Sears catalog gives us an excellent idea. Note that a size 14 is is a very small size, with a waist measurement of 27 inches!

Chapter 1

Gowns and Dresses

Neat Stripes in 80-square percale, set off with bold bands and belt of matching color. $2.79 [$35-45] **Broadcloth Coat Dress** with smart shawl collar, firmly finished buttonholes. $2.79 [$35-45] **Bright Print** with Ruffles. $2.79 [$35-45] **Percale Print** dashed with black. Stunning diagonal shoulder yoke and skirt border. $2.79 [$35-45]

Note: The market for the semi-formal and "dressy" dresses from the early 1950s is stronger than the market for cotton day dresses. Collectors worry about looking "frumpy" in these clothes, but there is nothing quite like them for summer comfort. Paired with "wedge style" sandals and cotton socks, these dresses are a very sophisticated fashion statement.

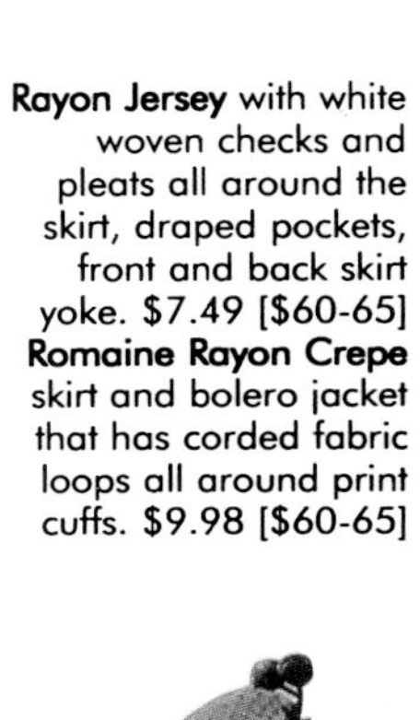

Rayon Crepe with white polka dots, floating panel dress. $6.39 [$40-45] **Print with Black Mesh** design and colorful flowers. $6.98 [$40-45]

Rayon Jersey with white woven checks and pleats all around the skirt, draped pockets, front and back skirt yoke. $7.49 [$60-65] **Romaine Rayon Crepe** skirt and bolero jacket that has corded fabric loops all around print cuffs. $9.98 [$60-65]

Rayon Alpaca Skirt with rayon crepe top, bolero and cummerbund belt. $8.98 [$70-75]

Rayon Shantung. Versatile suit-dress to wear with or without a blouse. $8.69 [$65-70]

Spring/Summer 1950

Surplice Neckline, skirt with front over drape and big bow. Rayon taffeta. $6.98 [$60-65] **Removable Collar and Cuffs** of white, fur-like fabric, full-circle skirt. Rayon taffeta. $9.69 [$60-65] **Front-draped Neckline,** full skirt. Ribbed rayon crepe. $9.49 [$70-75]

Fall/Winter 1951

Mitered Stripes with white broadcloth revers and button panel. $1.99 [$30-35] **Scroll Appliqué** on crisp white piqué revers, printed percale. $2.49 [$35-40] **Checked Adjustable Tie-back** fits any figure. Yoke and bias pockets with wide contrast bands. $2.59 [$30-35]

Checked Zipper-front with wide bias trim at rounded revers and cap sleeves. $2.39 [$35-40] **Closely Set Buttons** avoid gapping; set-in sleeves, pointed rolled collar, sweetheart neckline. $2.69 [$35-40] **Striped Zipper-front** has contrast yoke, striped appliqué tulip at shoulder. $2.69 [$35-40]

Plaid Button-front has double-breasted effect, gored skirt, plastic belt. $2.99 [$30-35] **Zipper-front Novelty Stripe** with solid yoke and border front. $2.99 [$35-40] **Mandarin Collar** button-front in multicolor paisley print. $2.99 [$30-35]

Striped Zipper-front with solid tunnel collar, curved cuffs, and double-cone pocket. $2.99 [$35-40] **Buttons for Trimming** on white waffle piqué collar and pocket flap accent bright novelty print. $2.99 [$35-40]

Bodice of Alencon-type rayon lace studded with sparkling rhinestones, American beauty rose, and a separate net stole. The skirt is two layers of net of rayon taffeta. $15.98 [$85-95] **Heavy Slipper Satin** bodice, satin skirt, and slip-on gloves of exquisite Alencon-type rayon lace for a one-piece evening dress. $15.98 [$85-90] **Rayon Faille Sheath Dress** for early and late evening parties. $10.98 [$65-75]

Tucked Casual in rayon gabardine, perfect for school or office with finely stitched sunburst tucks on bodice and skirt front, set-in pockets, rhinestone buttons, black belt. $6.98 [$55-60] **Suit Dress** with medallion pin of gold-color metal and grosgrain ribbons. $8.49 [$55-60] **Our Finest Gingham** woven plaid, $7.79 [$35-40] **Finest Quality Waffle Piqué** with woven check gingham collar, cuffs and pocket flaps. $8.19 [$40-45]

Pinwale Corduroy. Metal buttons and chain, four-gore bias skirt. $6.98 [$25-30] **Moire Taffeta.** Self-patterned rayon, gold-color metal chain and buttons, big cuffed pockets. $5.59 [$25-30] **Junior-style Dress** for proms and parties, our finest rayon taffeta, full-circle skirt, wonderful for dancing. Simulated pearls included. $5.98 [$25-30]

Bodice and Piping on separate jacket are plaid rayon taffeta, jewel-neck, pleated skirt. $6.98 [$30-35] **Contrast-color Appliqué** and binding, set-in pockets, reinforced belt. $6.98 [$40-45] **Checked Spun Rayon** dress and jacket. Bias bands on bodice. $6.98 [$30-35]

Slenderizing White Stripes and printed design coat dress, front and back shoulder yokes, plastic belt. $3.98 [$40-45] **Young, Easy-to-wear Dress,** polka dots; slim skirt has 1-gore front, 3-gore back. $5.79 [$55-60] **Faille-ribbed Rayon** with detachable bows, cuffs, and over collar of checked rayon taffeta. $6.29 [$55-60] **Frosty Shades** with floral fabric appliqué. $5.98 [$65-75]

Big Collar Dress of rayon gabardine, collar pointed in back. $7.90 [$55-65] **Coat Dress** of rayon gabardine with zipper all the way down the fly-front, braid embroidery on the pocket. $7.90 [$55-65] **Tucked Pocket Dress,** rayon gabardine, pockets across the front give a peplum effect. $7.90 [$55-65]

Bottom left to top right: **White Brocaded Satin** with crisp black taffeta, all rayon, silver-color metal chain and buttons. $6.98 [$55-60] **Shadow Plaid** spun rayon, smart coat dress with shoulder gathers for flattering bustline. $6.98 [$40-45] **Rayon Crepe** dress with flattering deep neckline, big bow, dirndl skirt. $3.98 [$40-45] **Spun Rayon Dress,** applied design on hip pockets that looks like embroidery. $4.98 [$45-50] **White Bodice Front** with checks that match skirt, deep shoulder flanges. $5.98 [$50-55]

Top left to bottom right: **Spun Rayon** with white checks, shoulder flanges for bustline fullness. $3.98 [$40-45] **Flattering Draped Neckline.** $4.98 [$45-50] **Rayon Crepe** with white polka dots. $3.98 [$40-45] **Print Dress** with shoulder pleats, shoulder bow, front peplum. $3.98 [$50-55] **Spun Rayon** with bright trim around shoulders, four-gore bias skirt. $5.98 [$55-65]

Youthful Black Dress. Skirt has front over drape and a big bow that do wonderful things for your hipline. Wing collar has a scatter pin of sparkling rhinestones. $8.98 [$70-75] **Rayon Alpaca** with print, easy-to-get-into coat dress with shoulder flanges for bustline fullness; stitched pleats in skirt front. $8.39 [$50-55] **Deep Tucks on Bodice** front, hard-to-find long sleeves. Bands go all around the skirt, dip to a V in back. $10.49 [$75-80] **A Suit Dress** that's young looking, easy to wear, and extremely slenderizing. $8.98 [$65-70]

Tailored Suit Dress with shoulder flanges and detachable cuffs and over collar of white cotton piqué. $8.98 [$55-60] **Ribbed Rayon Crepe** looks like silk for a youthful dress with expensive details: rhinestone clips, shirred shoulders, fine tucks, pointed gores in front and back of skirt. $12.98 [$65-70] **Fine-ribbed Crepe.** Front yoke of beautiful cotton lace, dyed to match. $9.98 [$75-80] **One-piece Tunic Dress** with sleeve insets and bands of matching rayon satin. $8.98 [$75-80]

Corduroy Two-piece Suit gives you a three-way jacket: belted, half-belted, or without belt. Stunning metal buttons. $12.98 [$55-60] **Corduroy Skirt** with unusual chevron pockets. $6.98 [$40-45] **Corduroy Slacks** with zipper, belt loops, front pleats, side pocket. $6.29 [$40-45] **Corduroy Weskit** with metal buttons. $3.98 [$30-35] **Corduroy Jumper** has plunging collar, metal buttons to the hemline. $8.95 [$40-45]

Windowpane Check Button-front uses solid color accent on appliquéd revers, square patch pocket, and cap sleeves. $2.98 [$30-40] **Scatter-Dot Dress** with deep round neckline front and back, perk-out cap shoulders, waffle piqué trim, white plastic belt. $2.98 [$30-40] **Two-tone Stripe** buttons to waistline. White piqué mandarin collar, cap sleeves, simulated pockets. $2.98 [$30-40]

Miracle Fabric, 100 percent nylon: youthful, two-piece dress with big collar, shoulder flanges, easy-fitting peplum, set-in pockets. $11.98 [$50-55] **Pointed Bodice** set on to yoke for becoming bustline, shirred skirt, fabric loops and buttons, plastic belt. $9.98 [$40-45] **Lined Collar and Cuffs** to keep their shape, pleats are lined at top to make them stand out, rhinestone buttons, velvet belt. $11.98 [$40-45]

Diamond Print has tie-back belt, lacy medallions on yoke. $1.99 [$30-40] **Side-Buttoned Check Plaid** scoop-neck style, white bias binding, curved pocket. $1.99 [$30-40] **Striped with Double Revers** accented with white. $1.99 [$30-40] **Floral Print** with bands of white binding from shoulder to hipline, simulated pockets. $1.99 [$30-40]

Scalloped Schiffli Embroidery pointed front and back at neckline, forms skirt panel; drawstring shoulders. $5.59 [$30-35] **Plaid Broadcloth** loop-buttoned to below waist, curved pouch pockets, set-in sleeves. $3.98 [$25-30] **Chambray with Waffle Piqué** lattice work at cap sleeves and square neck, matching collar, side zipper, pearl-like buttons. $5.59 [$25-30]

Calico Border Print square-dance style with puff sleeves, low neckline, dancing ruffle. $3.98 [$25-30] **Broadcloth** with checked gingham vestee and kick pleats, buttons point to slim waist, crisp cuffs. $5.59 [$30-35]

Satin-stripe Plaid Chambray buttons to below waist, saw-tooth neckline. $3.98 [$25-30] **Broadcloth with Slimming Strip** of waffle piqué, soft gathers at waist. $3.69 [$30-35] **Schiffli Embroidery** on dark bands accents check gingham, hipline. $5.59 [$30-35]

Border Print Brunch Coat. Oriental print effect forms skirt and trimming, side-tied wrap. $2.98 [$40-45] **Plisse Brunch Coat** with white piping and smocking on big pocket. Double-breasted effect with tie-back. $2.98 [$35-40]

Tattersall Checks with contrast buttons and plastic belt. $2.99 [$25-30] **Novelty Print with White Piqué.** Buttons to below waist for easy step-in, button-trimmed tabs on collar, plastic belt. $2.99 [$30-35] **Striped Zipper-Front** with solid bands in diamond pattern. $2.99 [$30-35] **Lace Daisy Dress.** Broadcloth with folded bands of white waffle piqué. $2.99 [$30-35]

Rayon Shantung sleeveless dress, mandarin collar, stitched tucks on bodice, unpressed pleats around skirt, plastic belt and buttons. $5.49 [$30-35] **Rayon Shantung Taffeta** in a glamorous new coat dress, big collar, big black plastic buttons. $6.98 [$35-40] **Linen-like Rayon.** Hand screen-painted lilies, tie back belt, button back. $4.98 [$70-75]

Woven-check Rayon Taffeta with stand-out cuffs. $4.98 [$40-45] **Rayon Crepe with White Polka Dots.** Detachable collar, cuffs, and pocket flaps of white cotton piqué. $3.98 [$45-50]

Woven Gingham, multicolor plaid, mercerized for soft luster. Two-piece dress, white piqué trim. $7.98 [$45-50] **Rich, Heavy Embroidery,** youthful mandarin collar. $7.98 [$65-70] **Combed Yarn Voile** woven to look as though it were hemstitched, pointed bodice set on to yoke. $7.98 [$40-45] **Chambray Dress** has flattering draped bustline accented by crisp white lace, well-cut bolero. $7.98 [$75-85]

Short Formal for Daytime or evening, 100 percent nylon net with applied design, boned bodice, Bodice lining and attached underskirt are rayon taffeta. Long sleeved bolero makes complete ensemble. $13.98 [$75-95] **Dramatic Tunic Dress** with front yoke of net, cuffed bodice. $11.98 [$65-80] **Alluring Coat Dress** of pure silk. $12.98 [$55-60]

Eyelet Embroidered Batiste with becoming draped bodice. $10.98 [$50-55]

Corded Stripes and Solid Chambray go together with big pockets, smart cuffs. $3.87 [$30-35] **Figured Stripe Chambray** with multicolor embroidered look, loop buttoned to below waist. $3.87 [$30-35] **Colorful Plaid Gingham** with white piqué piping, pointed button-down pockets, and bias cuff bands. $3.87 [$25-30] **Contrast Cotton Fringe** around curved collar and pocket flaps, buttons to the hem. $3.87 [$25-30]

Lace Trimming on Broadcloth. Collar and pockets edged with crochet-type lace and white bands. $2.99 [$30-35] **Novelty Dot-and-stripe Print,** pointed tabs with button trimming, set-in sleeves, plastic belt. $2.99 [$30-35] **Zipper Redingote Effect** in tattersall check, solid panel revers, cuff effect, daisy lace medallions. $2.99 [$30-35] **Gingham-like Plaid Print.** Double-breasted coachman style, split collar, pointed neckline. $2.99 [$30-35]

Woven Check Menswear Suiting. Side-button coat dress with big lapel collar. $8.49 [$45-50] **Flattering to Full Figures.** Shoulder flanges, four-gore bias skirt, scallops on bodice and pockets. $7.69 [$50-55] **Gabardine Button-front Coat Dress** with fine tailoring and becoming neckline. $6.98 [$45-50] **Shadow Plaid Spun Rayon** and acetate coat dress. $6.98 [$35-40]

Above: **Woven Check Taffeta.** Rever collar with velvet inset, velvet belt, rhinestone and plastic button trim. $6.98 [$45-50]
Left: **Fine Ribbed Crepe** easy-fitting jacket with braid embroidery. $10.49 [$75-80]

Tissue Faille Dress with front and back lace yoke, lined with net, tucks on front of bodice. Flower included. $10.98 [$65-70] **Fine Ribbed Crepe** with matching bodice of acetate taffeta. Bolero is flattering to hips. $9.98 [$50-55]

100 Percent Nylon, one of the popular "miracle fabrics." Has sparkling buttons, velvet belt, pleated skirt front. $8.98 [$40-45] **Dressy Type Suit** that's so smart for weddings and very special parties. Short, stiffened peplum, collar and cuffs of velvet with rhinestones and simulated pearls. $12.98 [$75-80] **Dressy Dress** has a front and a back yoke of dyed-to-match lace, lined in flesh color net and glittering rhinestones around the neckline and on the front yoke. $9.98 [$65-70]

Note: Although the market for vintage maternity clothes is limited, there are always a few collectors and designers looking for these items. The jackets look nice with leggings.

Maternity Dress and Jacket. Smooth rayon crepe with polka dots, jacket of faille-ribbed acetate and rayon. $7.98 [$40-45] **Maternity Suit** of woven plaid menswear suiting acetate and rayon. $8.98 [$50-55]

Quilted Cotton with full-circle skirt and low neckline. $6.98 [$50-55]

Nylon Net with satin bodice and brief bolero: a wonderful ensemble for weddings and formal parties. $15.98 [$75-80] **Wonderful Ensemble:** Bolero is quilted and studded with rhinestones; dress has halter bodice, low back, and a front midriff that gives you a very flattering bustline. $10.98 [$80-95]

Iridescent Taffeta with changeable two-tone effect. Alluring neckline, molded bodice, big skirt, rhinestone pins, attached crinoline petticoat. $8.98 [$60-65] **New "Peg-top" Silhouette,** around-the-hip pockets, shoulder tab with black velvet ribbon and rhinestone pin, velvet belt. $8.49 [$50-55] **100 Percent Nylon** dress with bias skirt, pleated all around, collar can be worn up or down. $9.98 [$50-55]

"Gibson Girl" Dress with fitted bodice, puffy sleeves, full skirt. $7.49 [$60-65] **Swishy Taffeta** trimmed with lovely white cotton lace, pretty stand-up collar, set-in pockets, fabric loops and buttons. $6.98 [$60-65] **Ottoman Faille** heavily ribbed rayon with cotton filling. Dramatic new coat dress with big plastic buttons, velvet collar, cuffs, and belt. $9.98 [$70-75]

Popular "Gibson Girl" Style with full sleeves, embroidered organdy collar, unpressed pleats all around skirt. $5.98 [$40-42] **Rayon Crepe with Polka Dots,** detachable collar of white piqué, "push-up" sleeves, fabric bow, plastic belt. $3.98 [$40-42] **Pinwale Corduroy** of sturdy, velvety-soft cotton. Collar, buttons, and bands on bodice are gold-color embossed cotton. $5.98 [$40-42]

Note: Locating all the pieces of a "wardrobe package" is rare, and the price quoted could be doubled for such a find.

Top row: **Sleeveless Stripe,** neckline cut low front and back. White appliqués on bodice and pocket. Wide white binding. $1.99 [$30-35] **Adjustable Tie-back** floral print. Smartly sleeveless with white piqué petal neckline piped in contrast. $1.99 [$30-35] **Zipper Front Check** with eyelet embroidery. $2.29 [$30-35] **Novelty Print,** white piqué notched collar with contrast braid. $2.59 [$30-35]
Left: **Colorful Plaid Step-in** buttons to below waist. Jaunty bow trims white band at bodice and skirt front, cap sleeves. $2.49 [$30-35]

Left: **Four-piece Wardrobe of Pima** cotton. $11.58 [$65-70]
Right: **Polka Dot** four-piece wardrobe. $6.48 [$65-70]

Hand-screen Border Print with gathered skirt and ribbon streamer belt. $3.98 [$30-35] **Dot and Solid Piqué** with two-piece look, patent-like plastic belt. $4.49 [$30-35] **Scroll Print Button-front** coat dress with decorative metal buttons. $2.98 [$30-35] **Lucky Horseshoe** twin print, even has a horseshoe neckline above its smart rounded yoke. $2.98 [$30-35]

Denim-like Chambray for tweedy texture, cotton coolness. Jacket has two collars, white cuffs too, simulated pearl ball buttons. $5.59 [$40-45] **Woven Stripe Seersucker** wears like iron. Buttons trim the pockets and cuffs of the semi-fitted jacket. $5.59 [$50-55] **Woven Check Suiting.** Small two-by-two buttons fasten the fitted jacket. Bias trimming, waffle piqué collar and cuff bands. $4.98 [$45-50] **Broadcloth with Blazer Stripe** waffle piqué trim, mitered at the new collar. Big buttons, neat-striped cuffs. $3.98 [$45-50]

Chevron Embossed Cotton with glossy finish, fitted at bodice, patch pockets and collar piped with white. $3.49 [$30-35] **Middy Torso.** Polished chambray with low-placed pleats, big gingham bow. $5.59 [$35-38] **Butterfly Print** with black accents in the binding, buttons at puckered shoulders, and shiny patent-like plastic belt. $3.98 [$40-42] **Polished Chambray Bareback** with halter neckline. Jet-like buttons march to the hem and trim pocket flaps. $5.59 [$40-42]

A Lovely Sheer with self "shadow" design, youthful double collar, pocket flaps, sparkling buttons, fabric loops. $9.98 [$45-50] **Cool Dress with Big Cape Collar** edged with dyed-to-match lace; fabric loops and buttons. $9.98 [$55-65] **Diamond and Swirl Design** dress has gathers and bow at neckline, 5-gore skirt. $8.98 [$55-65]

Sheer Handsome Basic Dress to wear casually or with your dressiest accessories. Pleats all around the skirt, satin belt, pearl-like buttons. $9.49 [$45-50] **Most Expensive** of our all-nylon fabrics: a coat dress made with gathers in skirt front, 2-gore back, set-in side pockets, plastic buttons, velvet band. $9.98 [$45-50] **Printed Puckered Nylon.** $8.98 [$50-55]

Short Evening Dress in nylon net with woven dots. Bare-top bodice, big billowy skirt has two attached underskirts, matching flowers, snap-on stole, shoulder straps (not shown). $15.98 [$65-75] **Fabric with Two-tone Effect,** alluring neckline, two rhinestone pins, full gathered skirt, crinoline petticoat. $8.98 [$55-60] **Wrap-around Coat Dress.** Molded surplice bodice, cinch belt, circle skirt. $8.49 [$55-60] **The Tiny Waist,** the billowy skirt, the completely feminine look! Woven stripe sheer, 50/50 silk/acetate. V-neckline in front and back, bow-trimmed sleeves, very full skirt. $9.49 [$45-50]

Coat Dress Has Ruffles and buttons all the way down the front, a full skirt that's shirred all around, tiny bow on mandarin collar, matching plastic belt, and pearl-like buttons. $9.98 [$35-40] **A Two-piece Dress** or, with a blouse, a suit. $8.98 [$50-55] **Dress with Pointed Bodice** for a flattering bustline, a skirt that's shirred all around.. $8 [$40-45]

Twin Print Coat Dress with flattering white ground print, cuffed pockets, four-gore skirt. $2.98 [$30-40]

Geometric Print accented with solid color. $2.98 [$30-40]

Lace Print with solid color bands and white cotton lace on the big revers and pocket. $3.49 [$30-40] **Striped Flower Print** with bias binding, pretty neckline, big pockets. $2.98 [$30-40] **Solid Color Coat Dress** with revers and inset vestee of white waffle piqué, embroidered braid trim. $2.98 [$30-40] **Zipper-front Dress** with black accents and white lace daisies. $2.98 [$30-40] **Slenderizing Stripes** accented with white waffle piqué revers and zigzag closing. $2.98 [$30-40]

WASHABLE SUITS

Woven Check Suiting is expertly tailored. Unlined jacket fits comfortably under a coat, white piqué collar and cuff bands are detachable. $4.98 [$45-50] **Two New Cottons:** the jacket is textured embossed cotton with check print; the skirt is denim-like chambray. $5.98 [$40-45]

Top left: **Flattering Dress** with detachable white piqué cuffs and dickey, rhinestone button trim. $6.98 [$40-45] ***Top right:*** **Our Finest Gabardine** in a coat dress with shoulder flanges, slim skirt, plastic belt and buttons, silk scarf. $7.69 [$40-45] ***Bottom left:*** **Gabardine Skirt and Bolero.** Woven check of wrinkle-resistant acetate and rayon. $6.39 [$40-45] ***Bottom right:*** **Smart Coat Dress** has washable under cuffs and collar of white piqué, set-in pockets, full skirt front, flared 4-gore skirt back. Black faille binding, black plastic belt and buttons. $8.49 [$35-40]

Three-pieces Make Four Outfits. Navy jacket and skirt of faille-type acetate and rayon; dress of navy rayon crepe with white dots. $12.49 [$55-70]

Nylon Lace. Skirt is shirred all around, yoke dips in back. Binding, bodice lining and attached underskirt are acetate taffeta. $12.98 [$65-70]

Antique Taffeta with a soft, silk-like luster, handsome matching braid embroidery, rich cotton lace medallions with metal nail heads. $10.69 [$75-80]

Long Torso Crepe of tissue faille. Yoke and gathers all around skirt. Narrow collar, detachable white faille over collar with rhinestones and lace. $9.98 [$50-55]

Beautiful Acetate Taffeta keeps its color and crisp new look. Tucked midriff, set-in vestee, attached taffeta petticoat with double ruffle of pink net. $9.49 [$75-80]

Woven Check Menswear Suiting. Tailored casual with button and buttonhole trim, wonderful big hip pockets, long back zipper. $7.49 [$40-45] **Rayon Suede,** a popular new fabric. Smart high neckline ties in back. $5.59 [$40-45] **Cotton Multicolor Stripes** with solid navy blue broadcloth, casual front yoke, button trim. $3.98 [$40-45]

Smart Printed Checks with waffle piqué collar and cuffs, jet-like buttons; full-shirred skirt. $4.29 [$35-40] **Polished Combed Chambray** dress with tucked yoke, rhinestone buttons, collar of white embossed cotton. $5.98 [$25-30] **Combed Gingham** dress with vestee effect of white waffle piqué, tucked and accented by pearl-like buttons. $5.59 [$35-40]

Extremely Smart Casual with button-down collar, detachable taffeta tie, gold-color metal buttons. $6.98 [$40-45] **Long Torso Silhouette** is accented by a middy cuff and pleats that go all around the skirt. Long back zipper; gold-color metal pins and chain with simulated jewels. $8.98 [$50-55] **Dress and Bolero.** Fitted bolero has dolman sleeves, convertible collar, big pockets, bright pin, and red silk scarf. Dress has sleeveless bodice, wide neckline. $8.98 [$65-70]

Long Torso Dress is wonderful for a Junior figure. Bodice is semi-sheer acetate and rayon crepe, has shoulder tab with rhinestone ornament. $8.98 [$50-55] **Princess-style Coat Dress** makes the most of a young figure. Gored in front and back for smooth fit. Buttons and piping are rayon velvet. $10.98 [$65-70] **Dress and Bolero** in taffeta. Rhinestone button trim on bodice of dress and sleeves of bolero. $9.98 [$65-70]

Nylon Net and Lace team up for alluring dress with shirred front bodice, net stole, and billowy skirt. New short evening-length skirt has lace ruffling all around and two attached underskirts. $17.98 [$85-95] **Nylon Net** with imported French lace makes a flattering dress for every age, every figure. Bodice is handsome Alencon-type rayon lace with attached camisole top of acetate taffeta. $14.98 [$85-95]

Chapter 2

Shirts, Skirts, Sweaters, and Pants

Roy Rogers and Trigger design on husky weight cotton with short sleeves. $1.59 [$45-55]

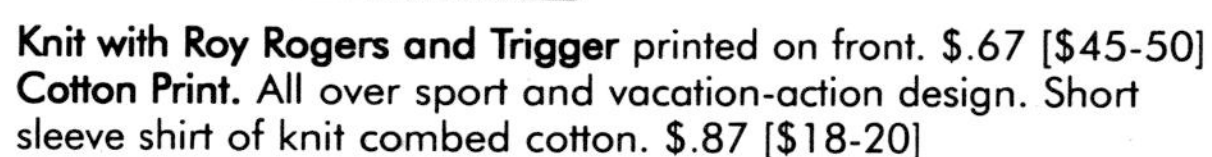

Knit with Roy Rogers and Trigger printed on front. $.67 [$45-50] **Cotton Print.** All over sport and vacation-action design. Short sleeve shirt of knit combed cotton. $.87 [$18-20]

Note: The Roy Rogers items crossover into the collectible marketplace. These items, made for little boys, are rare.

Just Right for "Dress-up" Wear. Saddle-stitched convertible collar, pocket flap, double yoke. $1.98 [$12-14]

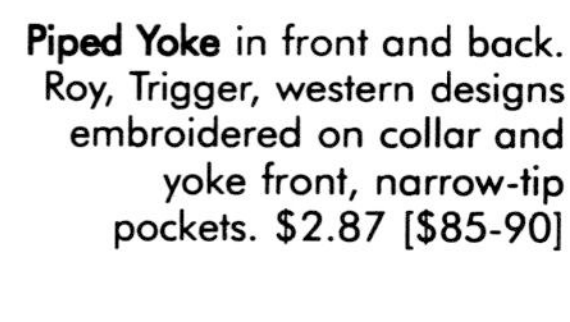

Piped Yoke in front and back. Roy, Trigger, western designs embroidered on collar and yoke front, narrow-tip pockets. $2.87 [$85-90]

Sporty Plaid. Loop closing for convertible collar. Pocket. Double yoke. $1.24 [$8-10] **Contrasting Yoke,** embroidered collar, chest pocket, two-button cuffs. $1.69 [$14-18] **Cotton Poplin** shirts. $1.24 [$8-10] **Cotton Broadcloth,** two-way collar, pocket, double yoke. $1.55 [$8-10]

T-shirt and Sweater Coat to match. Wear shirt alone when temperature soars. Wear sweater coat over it for cooler times. Cotton. T-shirt: $2.50 [$28-32] Sweater Coat: $3.50 [$32-35] Set: $5.50 [$60-70]

Raschel Knit; Chain Overlay Design. Sturdy cotton with a slightly raised overstitch for rough-textured look. $1.94 [$22-32] **Spear-fisherman Design.** Fine combed cotton, stretchy, sag resistant rib knit fabric. $1.94 [$22-32]

Three Pilgrims Swim Trunks. Cotton twill: $1.98 [$22-24] Satin: $2.98 [$22-24] All-nylon boxer: $3.98 [$22-24]

Pure Silk Print with orchid and bow pattern, johnny collar. $1.95 [$20-28] **Jewel Neckline** blouse (12 percent nylon, 88 percent rayon). $1.98 [$20-28] **Eyelet Embroidered** cotton blouse. Wear on or off shoulders. $1.98 [$15-20] **Cotton Batiste Blouse,** embroidered eyelet yoke. $1.79 [$15-20] **Cotton Broadcloth Midriff** with embroidered white eyelet ruffle. $1.59 [$15-20] **Rayon Crepe Blouse** with johnny collar. $.98 [$20-28]

Combed Cotton Two-Tone Knit with twin-layer fold-over front. $1.59 [$14-20] **Tailored Collar Knit** of selected full combed cotton. $1.45 [$8-10] **Heavier-weight Cotton** knit slip-over. $1.39 [$14-18]

Easy to Slip Into. Front and back of sturdy corduroy. Sleeves of panel-knit rayon/cotton fleece. $2.98 [$35-40] **Award-type Coat Sweater** for outdoor wear of 100 percent new wool (half worsted). $4.98 [$16-18] **Varsity Award Coat Sweater** is ideal for high school or club letters. Virgin wool worsted $6.35 [445-50]

Blazer Stripe Knit of fine combed cotton. Ribbed crew neck, cuffs. $1.49 [$8-10] **Practical Blazer Striped Knit** of full combed cotton. Ribbed crew neck, cuffs. $1.09 [$8-10]

V-neck Cable Knit Pullover of 100 percent worsted wool. $2.39 [$16-18] **Pullover with Nordic Pattern.** $2.98 [$22-28] **Sweater with Reindeer Pattern** on front and back. $3.97 [$32-38] **Ski Sweater** of worsted wool and cotton. $4.69 [$28-30]

Knit-in Winter Design Jacquard coat of 30 percent worsted wool, 70 percent cotton. Reindeer pattern on front and back. $2.97 [$15-20] **Knit-in Jacquard Slip-over,** worsted wool and cotton. Bruin and cub design. $2.97 [$15-20] **Heavy Weight Coat Sweater** knit of durable cotton. Varsity sleeve stripes. $1.97 [$12-15] **Genuine Knit-in Jacquard Coat** of sweater weight cotton. Rearing stallions on front and centered on back. $1.88 [$28-32]

Knitted-in Jacquard with leaping deer pattern on front. $2.97 [$28-32] **Norwegian-type** knit-in jacquard with skiers. $4.39 [$12-15]

Worsted Wool Pullover with self-knit cable pattern. $2.38 [$10-12] **Genuine Jacquard Knit** with striking ram's head pattern. $1.97 [$30-35]

King of the Cowboys: cotton chaps with picture of Roy and Trigger, cotton shirt, embossed leather belt and holster, cotton cowboy hat, toy pistol, lariat, and bandanna. $4.98 [$150-175] **Dale Evans Eight-Piece Cowgirl Suit.** Cotton skirt with print of Dale on front, vest, two-tone blouse, embossed leather belt, holster, hat, toy pistol, lariat. $4.98 [$150-175] **Brown and White Palomino Print.** Cotton chaps, flannel shirt, embossed leather belt and holster, toy pistol, cowboy hat, lariat, and bandanna. $3.98 [$65-70] **Girls' Eight-Piece Cowgirl Suit.** Palomino-print skirt and vest, plaid blouse, embossed leather belt and holster, toy pistol, hat, lariat. $3.98 [$65-70]

Action-packed Cotton Tweed Slipover with Roy and Trigger screen printed on front. $1.39 [$45-55]

Roy Rogers Cotton Broadcloth Shirt. $2.87 [$55-60] **Hawaiian Print.** Interlined loop-fastening collar, pocket flap, cuffs. $1.94 [$18-22] **Plaid and Plain Western.** Features whip-stitched collar. Pointed plaid front yoke. Interlined collar and cuffs. $1.94 [$12-15] **Printed Plaid** with interlined loop-fastening collar, tailored chest pocket. $1.59 [$8-10]

"Pony Skin" Printed cotton flannel shirt. Double-yoke construction. $1.67 [$15-18] **Flannel Plaid.** Double yoke construction. $1.89 [$8-10] **Flannel Plaid and Plain Western.** Piped double yoke, set-in dart pockets. $2.27 [$10-12] **Heavier Weight Flannel Plaid.** Double-yoke construction. $1.54 [$10-12]

Dressy Spun Rayon Gabardine shirts are washable. $3.95 [$32-38]

Classic Boxy Style. Tuck in or wear outside. Knit of cotton fibers. $.98 [$20-22] **Bold Plaid.** Fine combed cotton. $1.94 [$18-20] **Trim Multi-colored Pinstripes** on smart heather background. $1.94 [$18-20] **Jacquard Knit.** Fine combed cotton. Woven in pattern at front, back. Action-cut sleeve. $1.94 [$22-28] **Jacquard Knit Cardigan** with woven-in V-effect. Plastic buttons. Ribbed neck, cuffs, bottom. $2.89 [$28-32]

Multicolor Stripes go with all your skirts. Casual shirt style with pointed collar, buttoned cuffs on long full sleeves. Rayon crepe. $2.98 [$15-18] **Cotton Broadcloth Jacket Blouse.** Bat-wing sleeves, a contour-shaped waistband with buckle $2.98 [$18-20] **Three Collars.** A California fashion in fine rayon crepe. $3.49 [$22-28] **Bow-tie Collar** flirts with the neckline of your favorite suit, looks pretty without a coat. $3.79 [$22-24]

The Pleated Look, a mist of sheer nylon, full sleeves, young batwing collar, buttons in back. $7.59 [$30-35]

Sash Blouse of rayon tissue faille with smart wing sleeves, tucked-V—neckline. $4.98 [$32-38]

Dyed-to-match Lace in a flattering yoke effect. $2.29 [$28-32] **Handmade Look** with fine machine drawn-work, scroll embroidery. California-length double fabric sleeves. $4.98 [$32-35] **Peplum Blouse with Scallops** may be worn with peplum in or out. $3.98 [$28-32] **Worsted Wool Jersey.** Side-button collar changes completely when worn open. $5.98 [$28-32]

Triple Tabs and jeweled buttons on a turtle neckline with flange tucks at the shoulders. $3.98 [$28-32]

Cross Tucks on the front, collar, and sleeves. Fabric of rayon tissue faille, matching Alencon-type cotton lace. $5.98 [$22-28]

Thrify Acetates in fresh new patterns. Two-way sport collar has loop for top button; hemmed square bottom. $1.87 [$28-35]

Men's Short-sleeved Sport Shirts in cotton crinkle crepes that look and feel like seersucker. Two-way collars; square bottom, two pockets, gathered back. $1.87 [$32-35]

Economy solid-color knit cotton shirt. $.50 [$8-10] **Our Finest** solid-color knit cotton shirt. $.79 [$8-10] **Good** full-combed cotton blazer stripe knit. $.69 [$8-10] **Better** full-combed cotton novelty stripe knit. Heavier than "good" quality at left. $.87 [$8-10] **Best** combed cotton blazer stripe knit. Made of long staple yarns for extra smoothness, greater strength. $.99 [$8-10]

Short-sleeve Cotton Broadcloth. Blue, maize, or white. $1.23 [$8-10] Boys Crinkle Crepe Shirt. Red or blue. $1.37 [$12-15] **Broadcloth Mercerized.** Blue or white. $1.37. [$8-10]

Exclusive Roy Rogers cotton knit shirt with "day-glo" design. $.55 [$45-50] **Cotton Play Shirt** with knitted-in design. $1.15 [$8-10] **Full-combed Cotton Knit** shirt with action design. $.79 [$15-18] **Cotton Play Shirt** with knit-in real jacquard jungle design. $1 [$18-20]

Rib Interlock Knit cotton award coat sweater. $2.15 [$25-30] **Genuine Jacquard Knitted-in Design** cotton coat sweater. Wild pony designs on each side of front; large, twin pony heads centered on back. $1.94 [$35-40]

Three-color Club Stripe combed cotton knit. $.69 [$8-10] **Full-combed Cotton Knit** with set-in ribbed crew neck. Zebra design on front. $.67 [$12-15] **Genuine Links Knit** two-ply cotton. White with blue or green. $1 [$15-18]

Floral Print cotton broadcloth. $1.67 [$8-10]
All-acetate Rayon. $1.77 [$8-10]

Small Boys' Pajamas: White Cotton Broadcloth. Tightly woven: 172 threads per square inch. $1.89 [$8-10] **Short-sleeve Skip Dent Cotton Coat Pajamas.** $1.84 [$10-12] **Roy Rogers Short-sleeve** combed cotton knit pajamas. Ribbed pant cuffs. Fly front. Elastic boxer style waist. $1.95 [$45-50] **Cotton Broadcloth Coat Style.** $1.84 [$12-15]

High-count Cotton Broadcloth dress shirt. $1.49 [$10-12] Short-sleeve Printed Plaid Cotton Broadcloth. $1.27 [$8-10] **Cotton Broadcloth Western.** Vivid, glowing colors, yet completely wash. $1.74 [$12-15]

Roy Rogers fine cotton broadcloth shirt. Two slash pockets. Embroidered collar points. Looped cord yoke trim. $2.87 [$50-55] **Bike-type All Cowhide Belt.** Hip shaped "go-to-town" rodeo style with adjusting thong lacing. $1.98 [$45-50] **Official Roy Rogers Rodeo Hat.** 100 percent new wool felt with laced brim. Roy Rogers' name printed on band. $1.94 [$60-65]

Solid Color Cotton Poplin. $1.33 [$10-12] **Short-sleeve Better Printed** cotton crinkle crepe. $1.27 [$12-15] **Short-sleeve Brilliant Printed** rayon. Handsome all-over pattern. $1.40 [$12-15]

Best Quality Striped Pullover. Full-combed cotton. Knit tape neck finish prevents sagging, a feature usually found only in more expensive knit shirts. $1.25 [$8-10]

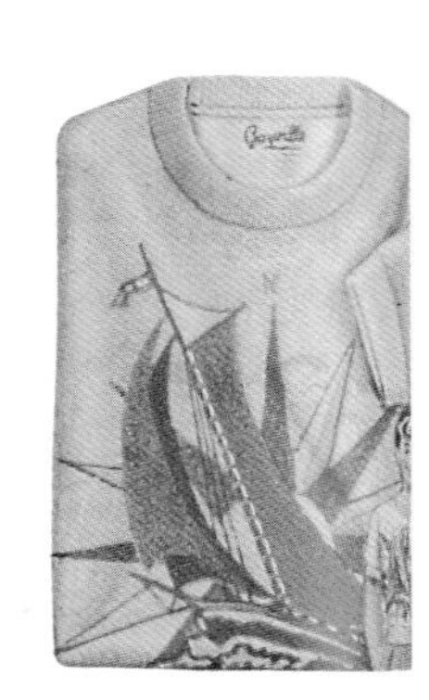

Better Full-combed Cotton novelty stripe knit. $.97 [$8-10] Cotton Knit Shirt with thunder bird design hand screened print. $.94 [$20-22] **Cotton Striped Knit Shirt** at a real low price! $.79 [$8-10] **Cotton Terry Knit Shirt** with "day-glo" design. Absorbent terry tops for sports, beach wear, and play. $1.23 [$22-24]

Hand Screen-printed cotton jacquard play shirt. A "natural" for boys. $1.29 [$20-22]

Mesh Cotton Knit with ombre panel shadow effect knitted-in. $1.39 [$15-20] **Cotton Play Shirt.** Jacquard knitted-in design. $1.59 [$15-20] **Cotton Shirt with Knitted-in Design.** $1.79 [$15-20]

Cotton Broadcloth Pajamas. Chest pocket. Trousers have two-button adjustable waist. $2.23 [$12-15] **Short-sleeve Cotton Knit Ski-style Pajamas.** Striped top and solid bottom. $2.17 [$12-15]

Cotton Broadcloth Shirts: Pineapple Pattern, rust on white. $1.97 [$35-40] **Plaid** in light gray, light blue, light green. $1.97 [$12-15] **Novelty Plaid,** brown, gray, green. $1.97 [$28-32] **Exceptional Quality for the Price.** White, light tan, light gray, light blue, light green. $1.97 [$12-15]

Cotton Crinkle Crepe Shirts: Mexican Hat. $1.97 [$35-40] **Bias Plaid, Leaf Pattern.** $1.97 [$22-24]

***Left to right:* Three-color Jacquard Knits:** Combed cotton. Reinforced neck, shoulder seams. **T-shirt.** Sailboat design. $2.25 [$28-32] **Coat.** Pattern compliments T-shirt. $3.20 [$32-35] **Set.** Maroon or blue. $4.95 [$60-70] **Knit Cotton Combinations:** Handsome tailored collar with emphasis on "neckline neatness." $1.74 [$15-18] **Newest in T-shirts** for casual or dress-up. $1.97 [$12-15] **All-time Summer Favorite.** Solid color body, contrasting tailored collar, two-button placket closure. $2.95 [$28-32]

Soft Combed Cotton in a variety of casual knits. Reinforced neck, shoulder seams: **Two-tone Link Pattern.** Bright top, darker bottom. Rib neck, sleeves, waist. $1.97 [$28-32] **Vivid Design.** $1.97 [$28-32] **Printed Scene** on Raschel fabric. Rib neck, waist, short sleeves. $1.97 [$35-40]

Soft Combed Cotton in a variety of casual knits. Reinforced neck, shoulder seams: **All-over Knit Pattern.** $1.74 [$28-32] **Cool Mesh Knits.** $1.74 [$15-18]

Rayon Gabardines in Eight Solid Colors. Gathered sleeves, shirred back, two-way sports collar. $2.95 [$32-38]

Spun Rayon Solids in choice of eight colors. Backs and sleeves shirred to give no-bind fullness, two-way sports collar. $2.95 [$28-32]

Double-printed Small Check Pattern cotton sports shirt of medium-weight cotton twill. $2.95 [$18-22]

Note: Full circle skirts were as popular for little girls as they were for women in the early 1950s. Print motifs included dancing dolls, florals of all kinds, poodles, geometrics, stripes, and plaids.

The Full-circle Skirt: Pretty Posies with elastic shirred waistline. $1.98 [$28-32] **Dancing Dolls** with sleek waistline, side zipper. Bright red braid belt trimmed with clever sewing spools ties in front bow. Multicolor print. $2.79 [$28-32]

Note: Vintage clothing collectors tend not to buy the 1950s blouses with ties at the neckline. If the bow can be removed, the shirt is more versatile. Often collectors wear the shirts with vintage gabardine wool suits and jackets, or dressy skirts of the period. Nylon is not as popular as rayon.

Tiny Pin-tucks and wider shell-edge tucks stripe this shirt of rayon tissue faille. Soft gold, white, navy blue, mint green. $4.49 [$22-24]

Elegant 100 Percent Nylon Tricot with dozens of novelty stitched tucks give a pleated effect. White, navy blue, soft pink. $5.95 [$22-24]

Lustrous Acetate Rayon Crepe with fine Val-type cotton lace insertion, tucking that looks hand-sewn, lace-ruffled jewel neckline. White, pastel pink. $2.98 [$22-24]

Dainty Suit Blouse of nylon tricot with fluffy pleated ruffles. Flattering bow neckline. Small pearl-like buttons. White, pastel pink, coral rose. $4.98 [$22-24]

Pleated Nylon Ruching in diamond pattern. Soft pink, white, navy blue. $3.98 [$24-28] **Little Dandy Look** of corded rayon and cotton bengaline, mandarin collar and deep curved yoke. White, soft pink, aqua blue. $1.98 [$22-24] **Panels of Daisy Patterned** embroidered organdy, dyed to match rayon tissue faille. Stunning in black, white, or soft pink. $2.98 [$28-32]

Dolman Sleeve Costume Blouse. Long sash ties to fit any waistline. Navy blue, coral red, white. $3.98 [$24-28] **Lace Hearts** on sheer Chantilly-type cotton lace yoke. Heart-shaped buttons, too. $1.98 [$22-24] **New Wavy Ripple.** Acetate rayon and nylon. White, soft pink, navy blue. $2.98 [$18-20]

Note: Flared and pleated skirts from the early 1950s are not as interesting to collectors as the straight skirts. During the early 1950s some wearers "pegged" the skirts, tapering the lines more tightly, and wore them with orlon or cashmere short-sleeve pullovers. Flared skirts with great detailing, such as buttons, hanging medallions, or big pockets tend to sell.

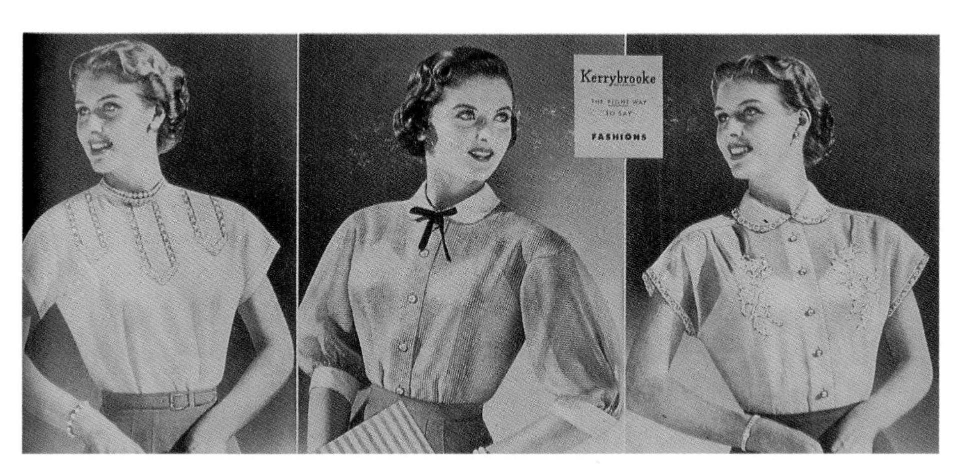

Jewel Neckline and Lace Wings. Rayon crepe with pointed panels of fine pin-tucks, inserted Val-type cotton lace, and heavy lace medallions in wing design. $1.98 [$24-28] **Young, Fresh, and Pinstriped,** a blouse to make you look like spring itself. Raglan shoulders, "chicken leg" sleeves, embossed-cotton collar and cuffs. $2.98 [$18-22] **Sheer Nylon with Embroidery.** Dyed-to-match lace on the round collar, buttons with rhinestones. $2.98 [$24-28]

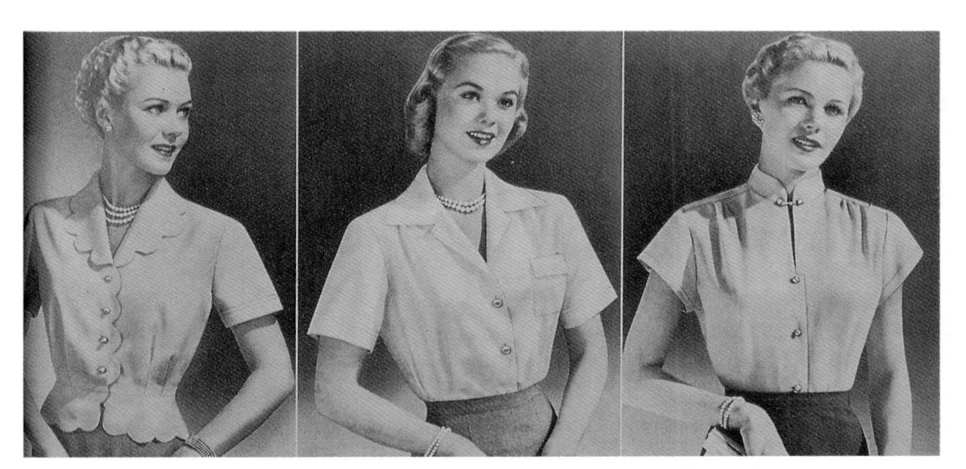

Scalloped Rayon Crepe Costume Blouse. Double fabric peplum. $3.98 [$22-24] Convertible Collar, dyed-to-match buttons. Shoulder pads. $1.98 [$22-24] **Texture of Linen** from 100 percent nylon. Pearl-like buttons, mandarin collar. $3.89 [$22-24]

Spring/Summer 1952

***Left:* Knife-pleated Skirt.** In Luscious Charcoal Gray. Princess-fitted waistline nipped by metal-tipped, leather-like cord through loops. Crisp trouser pleats. Charcoal gray. $4.98 [$18-20]
***Right:* Quilted Separates.** Whirling full-circle skirt teamed with dart-fitted vest in quilted cotton. Vest: $1.98 [$18-20] Skirt: $3.98 [$24-28]

Fall/Winter 1952

Wool Cardigan, only $2.94 [$20-22]

Top row: **Wool Jersey.** $2.98 [$24-28] **Nylon Pullover.** $2.94 [$24-28] **Jerkin and Pullover.** $3.74 [$28-32] *Bottom row:* **Jacquard-design** wool. $1.94 [$24-28] **Lustrous Rayon Boucle.** $3.87 [$22-24] **Angora/Wool.** $3.98 [$22-24]

Diamond pattern. $1.94 [$24-28] **Modified Turtleneck.** $1.98 [$22-24] **Two-tone Style.** $1.98 [$18-22] **Neat, Colorful Checks.** $1.98 [$22-24]

Fitted-style Polo Shirt. Modified dolman sleeves, multi-color stripes. $1.90 [$28-32]

Nubby Textured Cotton Poplin sport or dress-style shirt. White, tan, light blue, light green, gray. $1.97 [$24-28] **Pineapple Pattern.** $1.97 [$28-32]

***Above:* T-shirt.** Cotton terry cloth. $1.68 [$28-32] **Boxer Trunks.** Acetate twill. $2.97 [$28-32] **Set.** White, maize. $4.50 [$55-65] **Sport Shirt** with button front. $2.97 [$40-45] **Boxer Trunks.** Fully lined. Dark green. $2.97 [$18-20] **Set:** $5.75 [$55-65]

***Left:* Boxer Trunks: Cotton Twill,** tan, maroon, royal blue. $1.98 [$18-20] **Acetate Satin Twill,** maroon, royal blue, dark green. $2.97 [$18-20] **100 Percent Nylon,** medium green, maroon, gray, royal blue. $3.95 [$15-18] **Nylon Knit Brief.** Bright green, royal blue, red. $3.95 [$15-18]

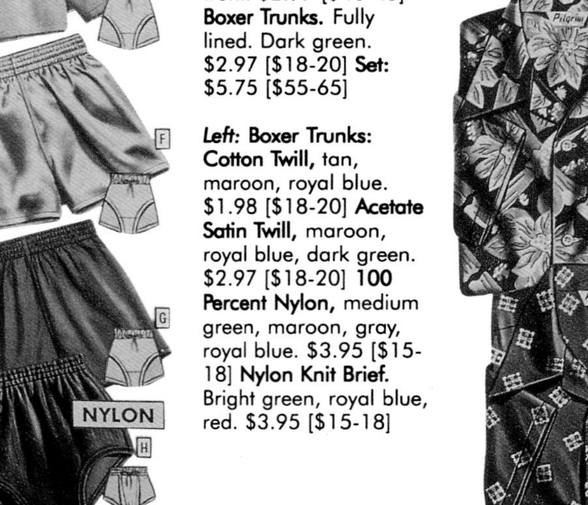

Rayon Crepe Print in gay floral pattern. $1.97 [$32-38] **Printed Rayon Crepe** in short sleeve Pilgrim sport shirt. Maroon, light gray, forest green. $1.97 [$32-38]

Cotton Terry T-shirts, Coats: Woven Plaid Fabric Collar and placket facing. $1.97 [$32-35] **Terry Jacket** with full zipper front and collar. White, medium blue. $2.95 [$32-35] **Cotton Terry Set** has bright sailboat scene. $3.95 set [$60-70 set] **Shirt Only:** $1.74 [$28-32] **Coat Only:** $2.47 [$32-38]

Full Circle Embossed Cotton colorfully striped in bright summer shades. $1.98 [$28-32] **Matching Sleeveless Embossed Cotton Blouse,** brilliantly striped in front to match the big whirl skirt. Solid white back. $1.19 [$20-22]

Cheerful Floral Print Broadcloth in a sweeping bedazzling full circle. Elastic shirring around waist, multicolor floral print. $1.98 [$28-32]

Plaid Broadcloth Circle cinched by a wide patent-like plastic belt. Big tunnel belt loops. $2.79 [$28-32]

Multi-plaid Whirl-away skirt in broadcloth with wide patent-like plastic belt, side zipper. $2.98 [$45-50] **Multi-colored Stripes** in denim. $3.49 [$45-50] **Colorful Flowers** bloom on crisp glossy embossed cotton. $3.98 [$55-65]

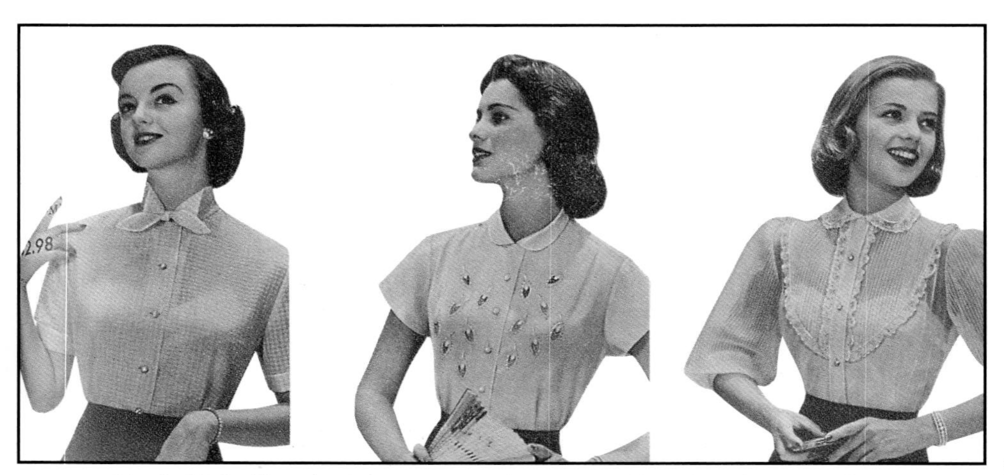

Wonderful Crinkle Nylon Fabric never needs ironing. Self-tie tabs at batwing collar. Sparkle buttons. $2.98 [$20-22] **Sheer Net Petals** set in with embroidery on acetate tissue faille. Mint green, white, pastel pink. $2.98 [$24-28] **Nylon Tricot Cloud-puff Sleeves** and yoke in permanent woven tucks. White, pink. $5.98 [$22-24]

Sweetheart Blouse wears a garland of cotton lace hearts. Yoke is permanent finish organdy in fine tucked effect. $2.39 [$22-24] **Nylon and Orlon:** two magic fibers in pin stripes so fine they almost look like solid color. Soft rose, white, soft gold. $2.98 [$20-22] **Nylon that Isn't Transparent.** Nylon ruching around collar, across front. White, pastel pink, pastel blue. $4.98 [$22-24]

Nylon with the Linen Look, batwing collar, rayon velvet tie, cameo-like tie pin. $2.98 [$22-24] **Sheer with Fine Stitched Tucks** across front, novelty stitching at shoulders. $4.98 [$20-22] **Shell-stitch Tucks** show up smartly with suits. $3.98 [$22-24]

Above: **Flaring Double Wing Collar** with posy 'n bow. Embossed cotton. White, maize, pink. $1.54 [$18-20] **Sleeveless Cotton Piqué** sports cross-tucked pockets, jaunty bow-tie. $1.69 [$20-22] **Vertical Tucks** topped with tabs and posy at collar. $1.89 [$22-24]
Right: **Sheer Nylon Sleeveless Camisole Blouse.** For those important spring-into-summer dates, let sleeveless nylon keep you looking cool and pretty. Trimmed with nylon lace. Posies and bow-tie at collar. Separate cotton batiste camisole with eyelet embroidered front. White, maize, pink. $2.74 [$22-24]

New Wool Worsted Jacquard-knit pullover. $3.89 [$38-40] **Jacquard-knit Coat Sweater**. Warm heavy-weight cotton. $2.29 [$20-22]

Long Sleeve Pullover. New wool worsted two-ply knit. $3.79 [$15-18] 100 Percent Australian Zephyr Wool. $4.69 [$15-18]

Fraternity Prep Award Coats, two-ply knit for extra warmth: Medium heavyweight. $4.79 [$40-45] **Heavyweight.** $5.49 [$45-50]

Cotton Pullover Sweater. Bright knit-in jacquard design. $1.95 [$32-35] **Two-tone Collar** model knit shirt of combed cotton. $1.49 [$32-35]

Wide Quilted Bottom dresses up this attractive four-gore cotton corduroy. Red, gold, purple. $5.98 [$35-40] **Big Crisp Box Pleats** all around, 50/50 wool/rayon in red and green plaid. $2.98 [$18-20] **High-notched Waistband** and unpressed pleats all around, 50/50 wool/rayon. Patent-like plastic belt. $3.98 [$18-20] **Heavily Quilted Full Circle** of shimmering acetate taffeta. The perfect glamour skirt for dates and parties. Navy blue. $4.98 [$40-45]

Our Finest shirt of washable 2-ply combed cotton. Short sleeves. $1.49 Long sleeves. $1.69 [$18-24]

Gold-color Crown Emblem on plastic belt sparks up this colorful cotton corduroy. Red, dark green, rust. $5.98 [$18-22] **Scalloped Pockets** with contrast piping on straight-line rayon and acetate check. Back zip and kick pleat. Navy check, brown check. $3.98 [$24-28] **All-wool Flannel** has huge patch pockets, pencil-slim skirt with inverted front pleat. $4.98 [$24-28] **Cotton Corduroy** in a big 3/4-circle sweep. Red, gold, purple. $5.98 [$28-32]

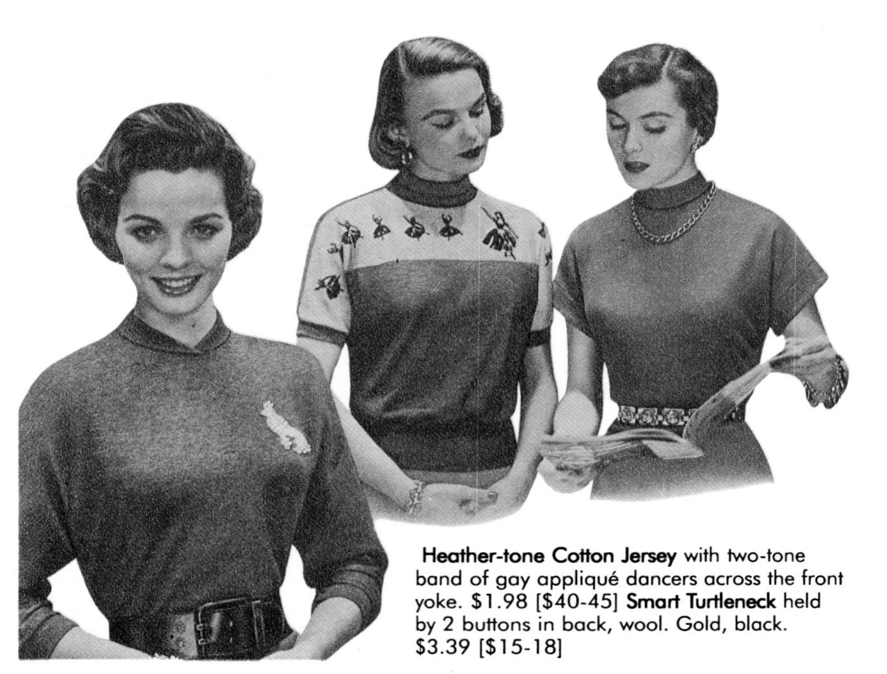

Heather-tone Cotton Jersey with two-tone band of gay appliqué dancers across the front yoke. $1.98 [$40-45] **Smart Turtleneck** held by 2 buttons in back, wool. Gold, black. $3.39 [$15-18]

Dolman-sleeve Cotton Jersey. Fifi appliqué is felt with sparkling eye. Heather gray, Heather gold, black. $2.98 [$40-45]

Sheer Nylon Tricot with lovely collar cupped in regal fashion. White, pastel blue, coral rose. $6.98 [$24-28]

Heavy Nylon Tricot, non-sheer, with convertible collar, simulated cuffs. White, gold, light blue. $3.98 [$20-22] **Tucked Front** with smocked effect at shoulders. Sparkling buttons. Pastel pink, white, light blue. $4.98 [$24-28]

Multiple Rows of Stitching make the collar roll smartly and the cuffs stand out. Honey beige, light blue, white, soft pink. $5.98 [$24-28]
Spaghetti Bows cluster at Peter Pan collar. Light blue, powder pink, white, powder gold. $4.98 [$20-24]

Raglan Sleeves gather softly into buttoned cuffs. Gold, bamboo beige, light blue, white. $6.98 [$24-28]

Long-sleeve Fitted Cardigan. Imported from England. $9.95 [$24-28] **Short-sleeve Fitted Pullover.** $6.95 [$24-28]

Jeweled Felt has a polished smoothness, 65/35 rayon/wool, king-size pocket set with simulated pearl studs. Medium gray, light navy blue, wine red. $7.98 [$65-75] **22-Gore Velveteen** is a real aristocrat. High quality cotton. Black. $8.98 [$40-45]

Three Yards Fuller than a circle skirt, crisp acetate taffeta with lots of fine gathers all around the waistband to accommodate your perkiest petticoats. Black, navy blue. $6.98 [$55-65]

Chapter 3

Sportswear and Swimwear

Denim Pants. Saddle pockets, wide belt loops, double seat, knee patch, lined crotch, taper legs. $3.89 [$25-30] **Gingham Shirt.** $2.79 [$18-20] **Stockmen's Style Pants.** 100 percent worsted wool. $13.98 [$30-32] **Two-tone Rayon Shirt.** $4.98 [$35-40] **Cavalry Twill Jodhpurs.** Leather knee patches. $4.49 [50-55] **Broadcloth Shirt.** $1.98 [$25-30] **Frontier Saddle Pants.** $4.49 [$50-55] **Rodeo Shirt.** $3.69 [$30-35]

Western-style Cowboy Suits. $1.89 and $3.98 [$40-45 each]

Cotton Coat Sweater. Fleeced inside. $.94 [$8-10] Virgin Wool. Hand embroidered. $1.98 [$8-10]

Two-piece Boxer Short Suits. $1 and $1.79 [$20-22] Cotton broadcloth shirt, solid color cotton poplin elastic waist shorts.. Tan, brown. Red, navy. $1.79 [$20-22]

Two-tone Corduroy Casual. For school, sports, or dress. $7.44 [$65-70] **Two-tone Casual Coat.** Wool and rayon. $5.95 [$65-75]

Woolen Shirt-style. $4.98 [$20-22] **Full Lined Mackinaw.** Wool. $8.75 [$20-22]

*Note: Pricing Dale Evans pieces is difficult. Complete sets in original boxes are the most valuable, so it is impossible to give individual prices, except for hats, belts, and shirts. Generally prices for individual items would be **$45-65**, and sets **$150-225**.*

Dale Evans Items: Twill jacket, twill shirt, five-piece set, six-piece outfit, short and long-sleeved polo shirts.

Dale Evans Accessories: Rodeo hat, western belt, western tie.

Rayon/Wool blend. Regular and large sizes. $$7.94-12.95 [$95-110]

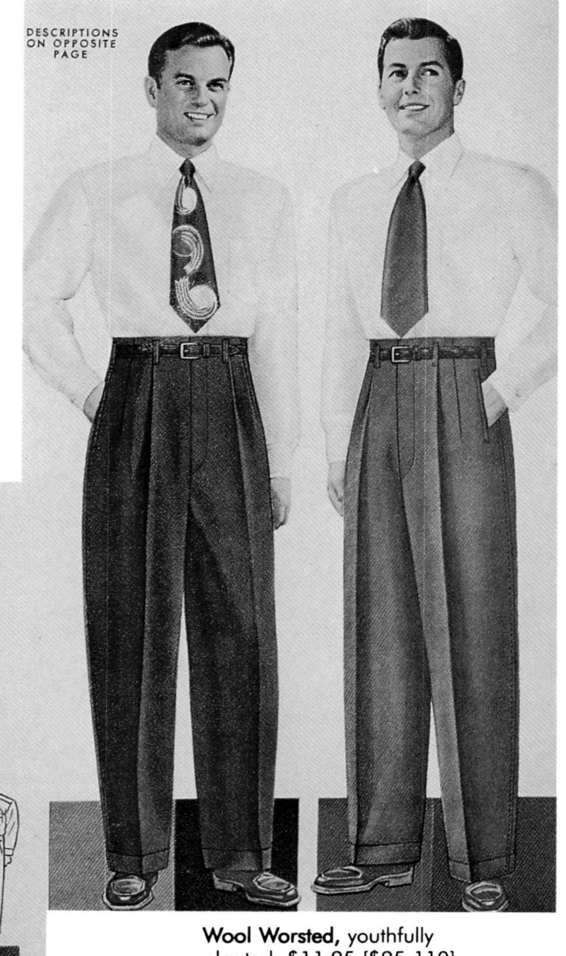

Wool Worsted, youthfully pleated. $11.95 [$95-110]

Note: The denims from the early 1950s which were bought for work-wear are rarely available on the vintage clothing market. They were simply worn out! Occasionally old stock vintage clothing becomes available including work clothes, overalls, or everyday "play" clothes. The market for old denim is specialized and unpredictable, and the prices quoted, although low, give a "safety net" for collectors who (like the author) do not deal in this market.

For the "King of the Cowboys." Hefty 9-oz. denim fights wear like Roy fights rustlers. Leather patch on back "branded" by Roy. Button fly: $2.05 [$65-70] Zipper fly: $2.17 [$65-70] **Extra Strong Construction.** Button fly or zipper fly. $1.69 [$30-35] **Preferred by Boys** who demand same strong construction as Dad's favorite overall. $1.94 [$45-50]

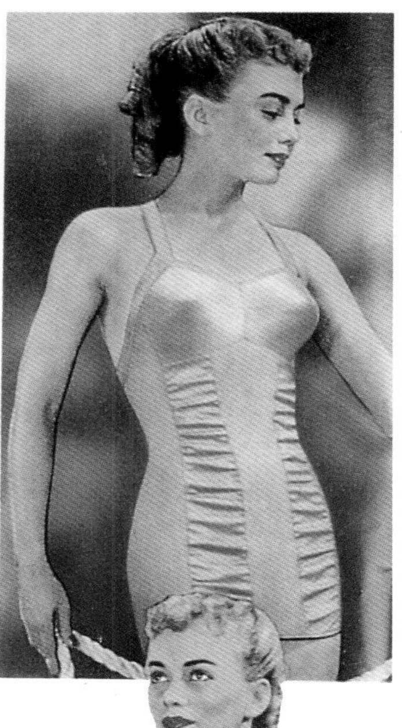

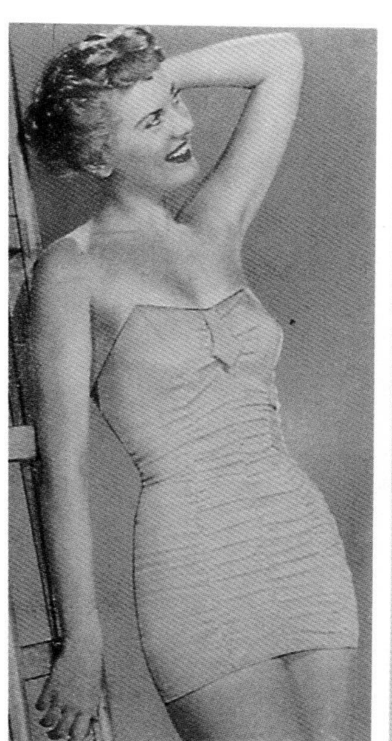

Kerrybrookes. Sears Exclusives. Made of acetate satin with latex yarn. $8.65-15.75 [$45-60]

***Above:* Shirring for Slimmer Lines.** Acetate jersey lined uplift bra with cuff top. Turquoise, lime, flamingo red. $5.95 [$50-55] **Appliquéd Butterflies** on rayon and cotton knit with Lastex yarn. Flamingo red, black, turquoise. $4.90 [$55-60] **One and Two-piece** shirred acetate satin swimsuits can be worn strapless. Blue, red, turquoise, lime. **Flared Skirt** makes hips and waist seem inches smaller. Low back, adjustable straps. $6.88 [$40-45] ***Left:* Shirring Nips in Waist,** flatters hipline. $8.85 [$45-50]

***Above:* Halter-top Sun Suit.** Cotton twill, patch pocket in back. $2.49 [$35-40] Camisole-top Sun Suit. Denim. Boyish cuffed shorts with back zipper. $2.98 [$40-45] Bra-top Sun Suit. Cord-stripe cotton. Buttoned halter straps. Shorts have side zipper. $1.98 [$50-55]
***Right:* Play Suit with Skirt.** Cotton sun suit for beach, sports or sunning. Add the button-front skirt and you're ready to go anywhere. $3.59 55-65]

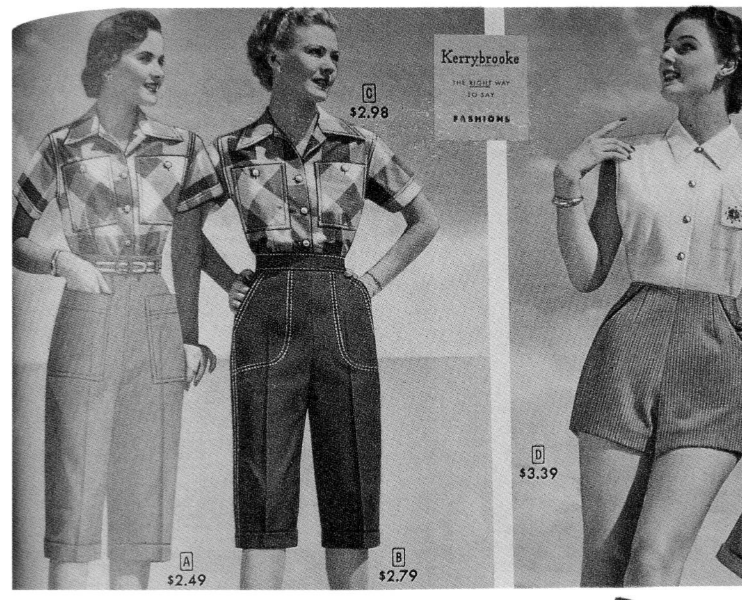

***Above:* Denim Pedal Pushers** tapered below knee. Novelty belt, two pockets, back zipper. $2.49 [$25-30]
Sailcloth Pedal Pushers stitched in white. $2.79 [$25-30]
***Right:* Assorted Shorts.** Corduroy, denim, cotton twill and linen-like rayon/cotton blend. $1.59-3.39 [$18-30]

Twill Slacks with pleat front. Good sturdy weight, man-tailored with zipper placket at side. $2.59 [$15-18] Regulation Length. $1.78 [$12-15] Bermuda Length. $1.88 [$12-15] Pedal Pusher. $1.98 [$25-30]

Men's Swimwear. Nylon, rayon, and cotton. $1.98-3.98 [$15-30]

Cotton Twill Rodeo Outfit with colorful jewel-like studs and silver-color nail heads. Slacks or jacket: $2.98 each [$55-60] **Boxer-style Blue Denims.** Jeans classic in true Western style $1.54 [$25-30] **Side-zip Denims.** For girls: $1.98 [$25-30] For teens: $2.29 [$25-30]

Denim in Two Weights or Cotton Twill. Metal-riveted front pockets. Back pocket. $2.49 [$30-35] **Eight-ounce Denim or Twill.** Tapered legs, pearl-like snap fasteners, and white braid trim on pockets. $2.69 [$30-35] **Lined with Cotton Plaid** flannel. Metal rivets reinforce two front pockets. $3.69 [$30-35] Shirt $2.49 [$15-18]

Four-piece Weekend Wardrobes. Linen-look cotton with white fringe or flowered. Four-gore flare skirt, sleeveless blouse, shorts, and camisole bra. Fringed: $8.98 set [$75-85 set] Flowered: $7.98 set [$75-85 set]

Pedal Pushers. Heavy weight twill: $3.49 [$25-30] Denim: $2.69 [$25-30] **Shorts.** Heavy twill: $2.79 [$18-20] Sailcloth: $2.59 [$18-20] Twill: $1.98 [$18-20] Denim: $1.79 [$18-20]

Our Best Kerrybrooke swim suits. $7.95-15.95 [$45-60]

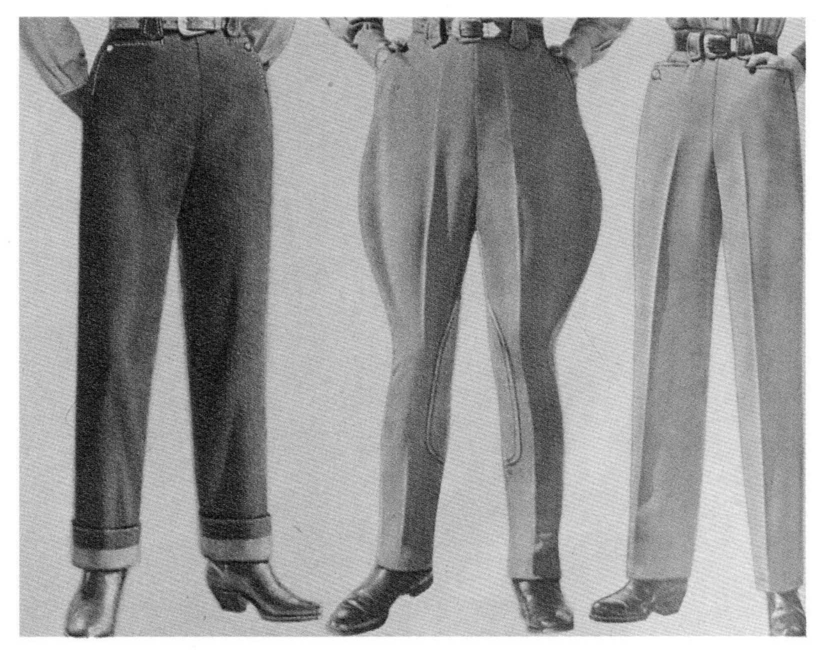

Denim Pipe-stem Pants. Pearl-like snap buttons at western-style pockets, chamois lined crotch. $3.98 [$30-32] **Cotton Cavalry Twill Jodhpurs.** Leather knee patches, wide western-style belt loops. $5.98 [$55-60] **Cavalry Twill.** Pearl-like pocket snaps, chamois-lined crotch. $4.98 [$35-40]

Combed Cotton Broadcloth Shirt. $2.98 [$10-12] Gingham, fine combed cotton. $2.98 [$10-12] **Combed Chambray Shirt.** $2.49 [$10-12]

Stripe and Solid Two-piece. $2.98 [$40-45]
Denim One-piece. $2.98 [$50-55]

Check and Solid. $3.98 [$50-55]
Broadcloth and Gingham. $4.39 [$40-45]

Play Suits. Solid with tattersall checks. $4.29 [$50-55] **Polka Dot.** $3.98 [$50-55]

Shirt and Slack Two-Piece Outfit. All rayon pleated front sharkskin effect slacks, cotton plisse print sport shirt. $5.75 outfit [$50-55 outfit] **Rayon Slacks,** rayon-print sport shirt. $5.75 outfit. [$50-55 outfit]

Authentic Western-styled Jeans and Jacket Outfits. Jeans: $1.98. Matching "Action Back" Jacket: $1.98 [$95-115 for set]

Denim Boxer Jeans. $1.35 and $1.12 [$15-18]

Sport Coat. Pinwale corduroy with husky shoulders, smooth and long-wearing rayon lining. $9.88 [$40-45] **Slacks.** Heavyweight, crease-resistant rayon and acetate gabardine. Pleated Hollywood styling. $4.65 [$40-45] **Sport Jacket.** Virgin wool "crash"-type tweed with decorative slubs of color, masculine padded shoulders. $15.75 [$40-45] **Slacks.** Virgin wool flannel, for casual or leisure wear. $7.95 [$40-45]

Sheen Gabardine Leisure Coat. Padded shoulders, rayon lining. $9.95 [$65-85] **100 Percent Virgin Wool** worsted sheen gabardine sport jacket. $18.50 [$95-115]

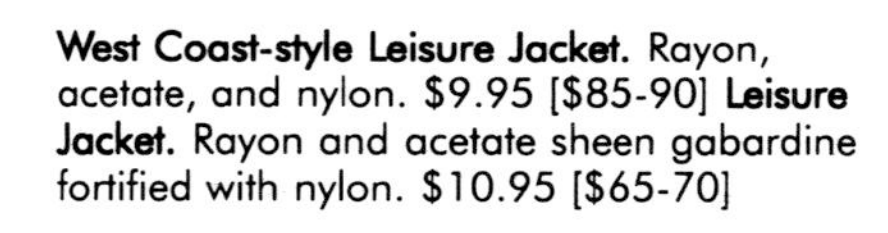

West Coast-style Leisure Jacket. Rayon, acetate, and nylon. $9.95 [$85-90] **Leisure Jacket.** Rayon and acetate sheen gabardine fortified with nylon. $10.95 [$65-70]

Chapter 4

Accessories

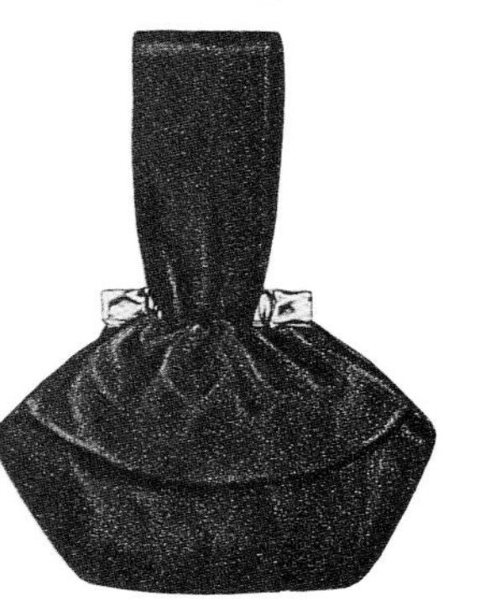

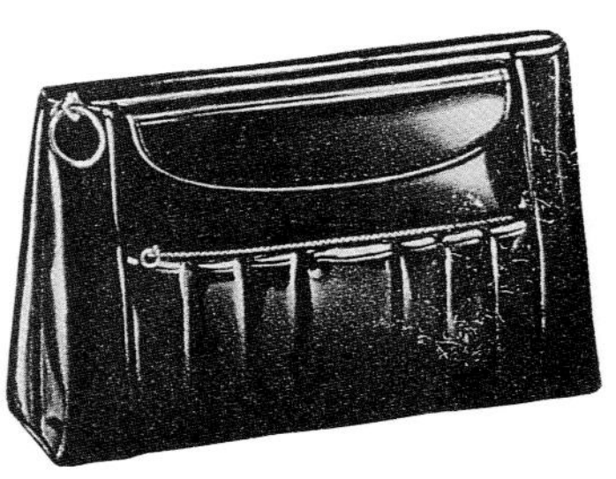

Clockwise from botoom left: **Underarm Zip-top** in patent-like or calf-like plastic. 12 x 7 inches. $1.99 [$15-18] **Pannier Handled Pouch** in rayon faille, Lucite trim, rayon lining, mirror. 11.25 x 6.25 inches. $1.99 [$30-40] **Pannier Pouch** in calf-grained plastic. 10.75 x 7.75. $1.99 [$30-40] **Adjustable Shoulder Strap.** 8 x 5 inches $1.99 [$20-25] **Pleated Rayon Faille Vanity.** Gold-color metal frame, Lucite clasp. 5 x 6.5 inches. $1.99 [$20-25] **Covered-frame Bag.** Gold-color metal clasp, rayon lining. 10 x 6 inches. $1.99 [$18-20] **Braided Beauty** in calf-grained plastic. 8 x 7.25 inches. $1.99 [$20-25] **The Box Bag** with wide top-handle in calf-grained plastic. 6.75 x 5 inches. $1.99 [$30-35]

Note: Pocketbooks of the early 1950s were designed in the most incredible shapes!

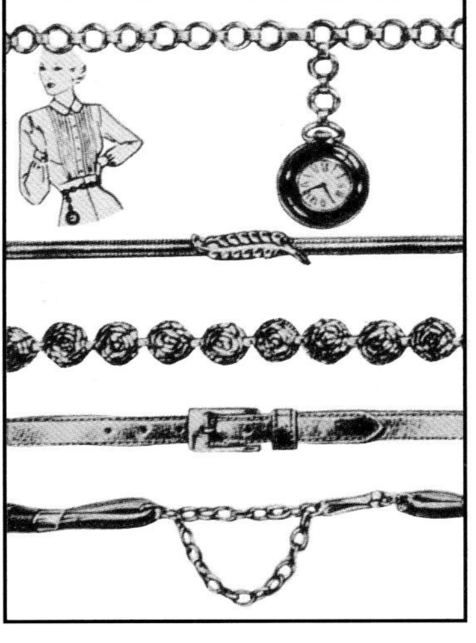

The Watch Belt, gold-color metal, enameled watch. $3.79 [$25-35] **Metal Coil Belt.** $.98 [$15-18] **Floral Chain Belt,** antique finished roses, metal. 1-inch wide. $.98 [$15-18] **Glittering Leather,** metal buckle, leather lined. $.98 [$20-25] **Dog Leash Belt,** plastic. $.98 [$15-18]

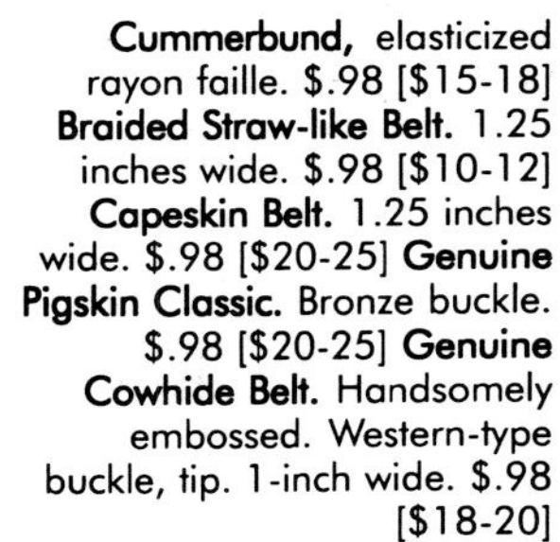

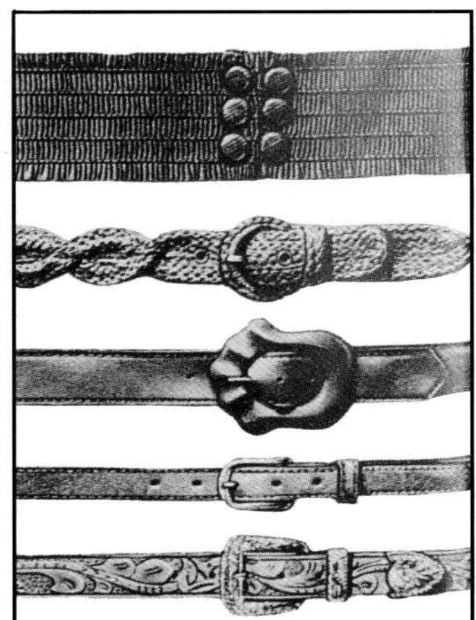

Cummerbund, elasticized rayon faille. $.98 [$15-18] **Braided Straw-like Belt.** 1.25 inches wide. $.98 [$10-12] **Capeskin Belt.** 1.25 inches wide. $.98 [$20-25] **Genuine Pigskin Classic.** Bronze buckle. $.98 [$20-25] **Genuine Cowhide Belt.** Handsomely embossed. Western-type buckle, tip. 1-inch wide. $.98 [$18-20]

Zipped Tophandler, mirror. 9.25 x 7 inches. $3.54 [$20-25] **Covered Frame Bag,** corded handles, mirror. 7.5 x 6.25 inches. $3.54 [$30-35] **Covered Frame Pouch.** Mirror. 8.5 x 6 inches. $3.54 [$30-35] **Popular Vanity Box.** Rayon lining. 6 x 6 inches. $3.54 [$45-55]

Note: Faille pocketbooks are very interesting to vintage pocketbook collectors. Often there is a full-view mirror inside the tops of the box styles. Faille holds up well, but sometimes shows wear in the corners and near the clasp. This wear lowers the value considerably. Dark shades such as black, navy blue, and brown are the most common colors found. The more unusual shapes tend to be the most expensive on the vintage market.

Rayon faille handbags, from left, clockwise: Shell Vanity, mirror. 8 x 5.75 inches. $3.54 [$25-30] **Braided Beauty.** Faille-tab, snap fastener, full-view mirror. 6.5 x 4.5 inches. $3.54 [$45-55] **Pleated Faille.** Double-corded top handle, Lucite clasp. 7.75 x 5 inches. $3.54 [$45-50] **Dressmaker Pouch** in softly shirred rayon faille. 9.5 x 5.5 inches. $3.54 [$40-45] **Double Corded Tophandler** with Lucite clasp. 10.5 x 5.5 inches. $3.54 [$40-45]

Note: Corde pocketbooks are very popular with collectors, some of whom collect only corde. The more intricate the work is, the more valuable, as a general rule. Raised and layered work is most desirable.

Popular Pouch in polished kip-calf. Mirror, coin-purse. 8 x 7.5 inches. $9 [$40-45] **Fashion Vanity** in polished kip-calf. Gold-color metal trims corners. Triple-cord handle. Coin-purse, mirror, rayon satin lining. 6.25 x 6.25 inches. $11 [$50-55] **Genuine Snakeskin** handsomely grained, fashioned into a dainty top-handle pouch with gold-color metal clasp. Mirror. 7 x 6.5 inches. $6 [$85-110] **New Look in Corde** with self-embroidered and quilted design. Zip-top, two side pockets, coin-purse, mirror. 11.5 x 5.75 inches. $9 [$95-120] **Box Bag Beauty** in glamorous corde-like rayon and cotton guimpe. Rayon and cotton cordette handle. Coin-purse, mirror. 7 x 5.5 inches. $6 [$40-45] **Softly Fashionable Faille** with a two-tier skirt. Lucite and gold-color metal clasp. Coin-purse, mirror. 11.25 x 7 inches. $6 [$40-45]

Note: Although collectors find the plastic pocketbooks fascinating; these pocketbooks do not bring prices as high as leather ones in similar styles.

Scallop-Top Underarm Bag, carefully tucked with smooth-running zip-top. Calf-like plastic. 12.5 x 5.75 inches. $1.99 [$20-25] **Tunic Pouch.** Fashionable dressmaker detail on front of rayon faille pannier handle bag. 10.5 x 6.5 inches. $1.99 [$25-30] **Box Bag** in plastic. Full view mirror. 7 x 4.5 inches. $1.99 [$20-25] **Collared Pouch.** Corded-top handle, plastic. 8 x 5.5 inches. $1.99 [$20-25] **Deftly Draped Pouch,** a stunning dress-up frame bag. 11 x 8 inches. $1.99 [$20-25] **Adjustable Shoulder-strap Bag** roomy, plastic. 10.5 x 5.75 inches. $1.99 [$20-25] **Music Roll Bag,** plastic. 8.75 x 4.25 inches. $1.99 [$20-25]

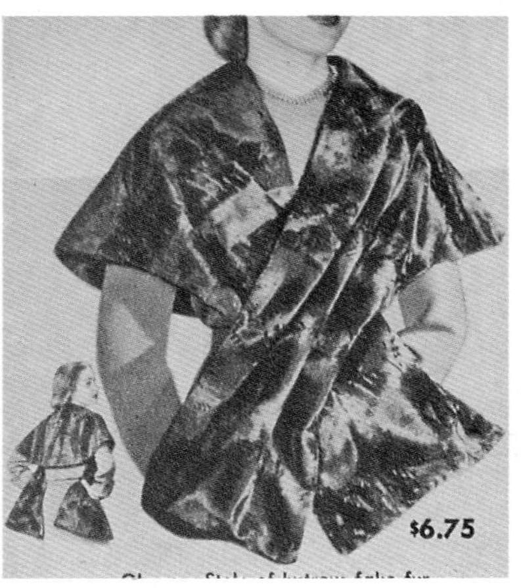

Glamour Stole of lustrous fake fur. Rayon and cotton cape stole looks like expensive broadtail. Rayon lining. Cape 13 inches Front 27.5 inches. $6.75 [$30-35]

Boxed-in Beauty. Corde. Smartly folded front, luxurious self-embroidered trim, turn-lock, coin purse, mirror. 9.25 x 4.75 inches. $9 [$80-85]

Silk Tassel Square. $.98 [$15-18] Gold-color crest on genuine leather belt. $1.98 [$20-22]

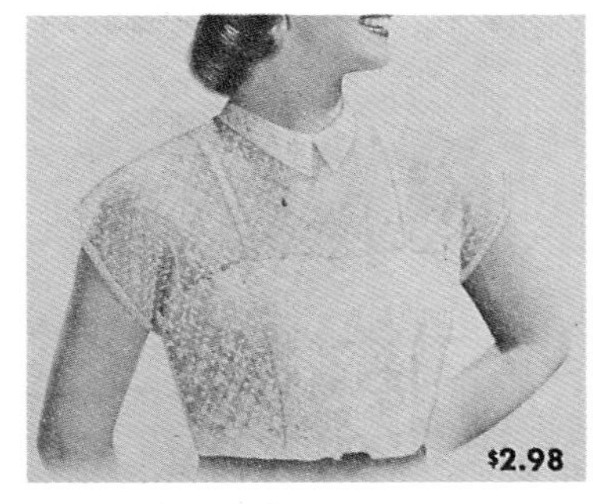

Lace Blouse of 64 percent acetate rayon, 36 percent miracle-wearing nylon, the glamorous fabric that is so rich-looking. Johnny collar. $2.98 [$25-35]

Glitter Duet: Gold-color trim and sparkling rhinestones on sheer silk chiffon triangular scarf. 9 x 36 inches. $1.98 [$15-18] **Gleaming** leather belt. $.98 [$20-25]

Note: Although shoulder-strap bags were available in the early 1950s, they were not as common as the pouch and box styles, with shorter handles.

Cowhide drawstring in keeping with the new casual look. 10 x 9 inches. $9.95 [$40-45]

The Vanity in genuine snakeskin. 7 x 5.25 inches. $6.50 [$75-85]

The Pouch in 100 percent nylon faille or sueded cotton. Coin purse. 10 x 6.5 inches. $5.78 [$30-35]

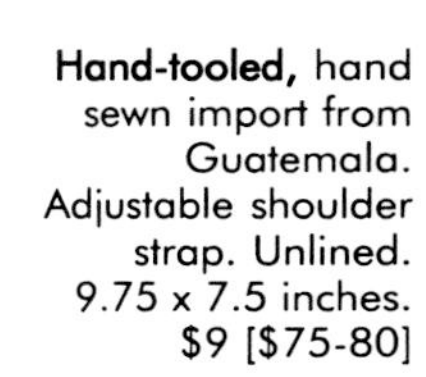

Hand-tooled, hand sewn import from Guatemala. Adjustable shoulder strap. Unlined. 9.75 x 7.5 inches. $9 [$75-80]

Queen of Hearts vanity in suede-like plastic. 7 x 5 inches. $3.59 [$25-30]

Silk twin-tipped square. It's a collar, a double point neckerchief, a three-petal tie, just to name a few ways to wear it. 17 inches square. $.98 [$18-20]

Box Bag in rayon faille. Gleaming gold-color compact on one end, many-purpose case on the other that is perfect for your change, cigarettes, etc. 7.25 x 2.5 inches. $3.59 [$40-45]

Above: **Silk Windsor Tie**, rainbow striped for added dash. 9 x 58 inches $.98 [$15-18] **Genuine Peccary Pigskin.** 10.75 inches long. $4.98 [$18-20] **Adjustable Shoulder-strap Bag and Belt** in top grain cowhide. Belt is jewelry touched with antiqued gold-color metal crest. $1.98 [$20-25] Bag is 9.5 x 5.25 inches $6 [$40-45]

Rayon Velvet Ballerina Cuff glove. 10.25 inches long. $1.59 [$18-22] **Pretty-as-can-be Pouch** in expensive-looking rayon velvet. 9 x 6 inches long. $3.59 [$45-50] **Twin Roses** bloom at your neckline for fashion-fresh flattery. In rich rayon and rayon velvet. 8.25 inches. $.98 [$18-20] **Contour Curve Belt** encircles your waist for a smart accent. In rich rayon velvet. $1.98 [$18-20]

Note: The beautiful leather handbags from the early 1950s are always a great bargain when compared with today's prices for leather pocketbooks. Highest prices are paid for unusual styles.

A: Vagabond Bag in polished saddle-finish cowhide. 8.75 x 4.75 inches. $6 [$45-50] ***B:* The Roomy Top-Handle Pouch.** In kip-calf. Wide-opening extension zipper top. 11 x 6.75 inches. Initials sold separately. $9 [$40-45] ***C:* Smart Pouch** in kip-calf. 8.25 x 6 inches. $9 [$45-50] ***D:* The Commuter** in handsomely grained leather. Zip-top. 10.75 x 7 inches. Initials sold separately. $6 [$25-30] ***E:* Fashionable Pannier Handle Pouch** in genuine corde. Luxurious self-embroidery on front. 10.5 x 6 inches $6 [$75-85] ***F:* Bumper-edge Adjustable Shoulder**-strap in kip-calf. All-around bumper, stylish loop handle, leather tab fastener. 10 x 6.5 inches. $9 [$40-45] ***G:* The Traveler.** 10.5 x 7 inches. Initials sold separately. $9 [$25-30] ***H:* Smartly Scalloped Genuine Corde Bag.** Attractive quilting on front of bag. Zip-top opening. 12 x 6.75 inches. $6 [$75-85] ***J:* All-time Best Seller** in top grade leather, spacious outside zipper pocket, two handy side pockets, inside zip pocket. 10.75 x 7.75. $7 [$25-30]

Note: Amazingly, a black plastic imitation of "expensive corde" was manufactured in the early 1950s. These "imitations" were also available in white and bone for summer use.

Drawstring Shoulder Strap has outside change pocket under flap. 8 x 8.5 inches. $3.79 [$35-40]

***Top row:* Yoke Vanity,** shirred in front and back. 8 x 5.5 inches. $3.97 [$45-50] Oval Bag. 8.75 x 4 inches. $3.59 [$45-50] Dressmaker Pouch, a covered frame bag with fashionable shirring. 9.5 x 6 inches. $3.97 [$40-45] ***Center row:* Sun Ray Bag,** a handy covered framed pouch with front pleating. $3.79 [$50-55] **Pannier Pouch** adds stylish draping. 9.5 x 6 inches. $3.97 [$50-55] **Scalloped Top-handler** with extended zip-top. 9.25 x 5.5 inches. $3.97 [$30-35] ***Bottom row:* Valise Vanity** has roomy two-section interior. 7 x 5.25 inches. $3.97 [$45-50] **Shirred Box** with smart corded handles. Flap-over top, metal turn-lock. 7 x 5.5 inches. $3.97 [$50-55] **Fetching Vagabond** equipped with large shirred pocket under flap-over top. 9 x 5.5 inches. $3.97 [$45-50] **Favored Drawstring Pouch** adds pretty petals on top. 8.25 x 8.5 inches. $3.79 [$40-45]

A: Pleated Drum Box in dressy rayon faille. Aglitter with gold-color metal frame, clasp. 5 x 5 inches. $2.69 [$55-60] **B: Sueded Rayon Vanity,** softly shirred. 7 x 5.75 inches. $3.59 [$40-45] **C: Double-decker Box.** Upper deck divided into two sections. Lower deck is a roomy galley. Leather-like plastic. 7 x 4.5 inches. $2.69 [$25-30] **D: Underarm Zip-top.** Calf-like plastic. 13 x 6.25 inches. Initials sold separately. $2.39 [$20-25] **E: Music Roll Bag,** calf-like plastic. 8 x 4 inches. $2.69 [$20-25] **F: Tear-Drop Pouch** in suede-like plastic. Lucite clasp. 10 x 6.5 inches. $2.69 [$20-25] **G: Royal Crest** (gold-color medallion) on adjustable shoulder-strap bag in calf-like plastic. 9.25 x 5.75 inches. $2.69 [$18-20] **H: Top-handle Swagger Bag** in leather-like plastic. 10 x 6 inches. Initials sold separately. $2.99 [$18-20] **J: Scallop-top Swagger Bag** in plastic, looks like expensive corde, garnished with luxurious quilting. 10.25 x 7 inches. $2.69 [$25-30] **K: Favored Pouch** in rich calf-like plastic with shirring with eye-catching tucks. 9.5 x 8 inches. [$25-30] **L: Drawstring Pouch,** suede-like plastic. Rayon lined. 7.5 x 8 inches. $2.39 [$20-25] **M: Initials.** Press-on initials. 18 karat gold finish with black enamel letters. Simply remove gauze back and press on bag. Size 1.5 x .5 inches. $.15 [NPA]

A: Ruffled Pouch in softly-sueded rayon. Smartly shirred. 10 x 5.5 inches. $2.69 [$30-35] **B: Triple-corded Barrel Bag** in calf-like plastic. 7 x 4.5 inches. $2.69 [$20-25] **C: Braided Beauty** with smart details. Inside zip-pocket. Rayon lined. Suede-like plastic. 8.5 x 5.25 inches. $2.69 [$25-30] **D: Capelet Pannier Pouch,** metal turn-lock. In rich, calf-like plastic. Rayon lined. 11.5 x 6.75 inches. $2.69 [$20-25] **E: Always Popular Box Bag.** Calf-like plastic. 6.75 x 5 inches. $2.39 [$25-30] **F: Tulip Flared Pouch,** petal-pretty bag in calf-like plastic. 12 x 8.25 inches. $2.99 [$18-20] **G: Adjustable Shoulder-strap Bag** in calf-like plastic. 10.25 x 6.75 inches. $3.59 [$18-20] **H: Top-handle Vanity Box** in suede-like plastic with triple tiers (front only). Rayon lined. 7 x 4.5 inches. $2.69 [$30-35] **J: Covergirl Hatbox Bag,** richly shirred. In smart rayon faille. 8 x 8 inches. $2.69 [$30-35] **K: Roomy Luggage Bag** in calf-like plastic. 9 x 6.75 inches. $2.99 [$20-22] **L: Key Finder,** a handy fitting for your bag. Attach keys to long chain that rewinds automatically in case. Plastic gift boxed. $1 [NPA] **M: Double-ette Caddy** keeps an eye on both your handbag and your gloves.

Handbag Dolls. These adorable figures, as cute as any toy, are shoulder-strap bags, too. In colorful leather-like plastic with padded bodice and arms. Zipper opening at back for bag compartment. Unlined. The Cowgirl and Pigtail Girl. 11 x 6.25 inches. $1.99 [$85-100] **"Dolly and Me" Set.** Zip-top Shoulder-strap bags in patent-like plastic for her and her doll. 6 x 4.25 and 4.25 x 2.25 inches. $1.19 [$40-45] **Bag 'n' Record Set.** Zip-top shoulder-strap bag in leather-like plastic, lined, packaged with nursery rhyme record that plays on a phonograph. Bag 6 x 4.5 inches. $1.87 [$35-40]

Upper left circle: **Bird Cage Bag** is this season's conversation stopper. Fashion-rich in glittering gold-color metal complete with its practical companion—a bright solid-color 18-inch silk square. $2.69 [$25-30] ***Top two:*** **Hold-everything Satchel** in Texon, a leather-like plastic. 8.5 x 6 inches. $3.97 [$18-20] Perky Box Bag in calf-like plastic. Handy mirror inside. 6.5 x 5 inches. $2.38 [$20-25] ***Center row:*** **Cute-as-a-button Pouch.** Three covered buttons and eye-catching calf-like plastic. 8.5 x 6.5 inches. $2.38 [$20-25] **Adjustable Set: Shoulder-strap** bag in calf-like plastic, 9.5 x 5.75 inches, and matching belt, $3.49 [$18-20] **Zip-top Clutch Bag** in corde-like plastic, embellished with luxurious self-quilting, embroidery. 8.5 x 5.5 inches. $1.27 [$18-20] **Double Tab Vanity** frame bag in calf-like plastic. 7.25 x 4.75 inches. $2.69 [$20-25] ***Bottom row:*** **Adjustable Shoulder-strap Bag.** Chain attached to 2 zippers for easy opening. 11.5 x 6.5 inches. $3.59 [$18-20] **Skirted Pannier Pouch** in rayon faille. 10 x 6.75 inches. $1.99 [$20-25] **Top-handle Swagger Bag** in long-wearing leather-like plastic. Rayon lined. 10 x 6 inches. $2.99 [$18-20]

A: Butterfly Pouch in rayon faille with sparkling gold-color highlights. Rayon lined. 8.5 x 7 inches. $2.69 [$40-45] **B: Big Box Bag** in calf-like plastic. Full-view mirror inside. 9.5 x 4.5 inches. $2.69 [$30-35] **C: Wonderful Vanity Box.** White linen-like plastic punctuated with all-over beautiful leaf design. 6.75 x 5 inches. $2.69 [$20-25] **D: Scallop-top Bag.** Imitation corde, self-quilting and embroidery, Zip-top. 10 x 7 inches. $2.69 [$20-22] **E: Lantern Box Bag.** Dressy rayon faille, full-view mirror. 7.25 x 4.25 inches. $2.78 [$40-45] **F: Adjustable Shoulder-strap Bag,** complete with plastic wallet and coin holder. In calf-like plastic. $2.69 [$30-35] **G: Unsnap** new converter for a top handle pouch. Rayon lined. 9 x 9 inches. $2.69 [NPA] **H: Underarm Shirred Bag** in calf-like plastic. 13.25 x 7 inches. $1.99 [$18-20] **J: Collared Pouch** with dressmaker pleats in calf-like plastic. 8 x 9.25 inches. $2.38 [$40-45] **K: Box Bag Success.** Calf-like plastic. 7.75 x 4.5 inches. $1.99 [$20-25] **L: Fashion Match Mates** in sueded rayon. Pleated box bag, 6.5 x 4 inches, and cuffed gloves, 9 inches long. $3.99 [$40-45]

Western Roll Bag and Belt in saddle leather. Gold-color metal horseshoe ornaments, adjustable shoulder-strap bag. $5 [$30-35] 9 x 5 inches. Matching contour belt. $1.98 [$18-20]

Tambourine Bag in Nylon Faille or Guimpe. Glitters with gold-color metal trim, rayon faille lining. Coin purse. Mirror. 7 x 7 inches. $5.89 [$45-55]

Carry the Covered Wagon Bag. Top grain cowhide, coin purse, mirror. 8.5 x 4 inches. $6 [$45-50]

2-in-1 Reptile Shoulder Bag converts to a drawstring pouch. Carefully stitched reptile square of precious snakeskin and lizagator. 9 x 9 inches. $9 [$55-65]

Genuine Corde Scallop-Top Bag. Enriched with self-embroidery and fashion-smart scalloping. 10.5 x 5.75 inches. $6 [$85-95]

Top row: **Drawstring Shoulder-strap Pouch.** Front flap-over pocket. 9.5 x 8.5 inches. $3.97 [$20-25] **Vital Vagabond Pouch** with scalloped flap-over top, unusual handle treatment. 9.75 x 4.5 inches. $3.79 [$40-45] **Drawstring Pouch** adds all-around ruffle. 9 x 9 inches. $3.59 [$40-45] **Center row: Perfect Pouch.** Smartly shirred, petal-pretty Lucite clasp. 8 x 6 inches. $3.59 [$40-45] **Terrific Vanity Box.** What style! 7 x 4.25 inches. $3.97 [$45-50] **Triple Frame Pouch.** 10.5 x 4.75 inches. $3.59 [$40-45] **Box Bag,** big enough for all your belongings. 9 x 4.25 inches. $3.97 [$45-50] ***Bottom row:* Vain Vanity** with arched front yoke. 8 x 5.25 inches. $3.59 [$50-55] **Button Box Bag.** Self-covered buttons, gold-color metal turn lock closing. 6 x 4.25 inches. $3.59 [$50-55] **Shirred Accents.** Dress-up version of a shoulder-strap pouch. 10.25 x 7.75 inches. $3.79 [$40-45]

Genuine Alligator. Made of an entire baby alligator. A fashion conversational piece. Lined with leather-like plastic. Made in Cuba. $16.98 [$75-85]

Note: Alligator pocketbooks from Cuba, which are made of an entire baby alligator, are rarely found in perfect condition. Many collectors prefer the more classic styles, which are usually leather lined and beautifully crafted. Twenty years ago the baby alligator pocketbooks were more popular than the classic (no head/paws) styles. When first imported from Cuba, these pocketbooks were offered at souvenir stands along the roads in Florida, as well as in the Sears catalog.

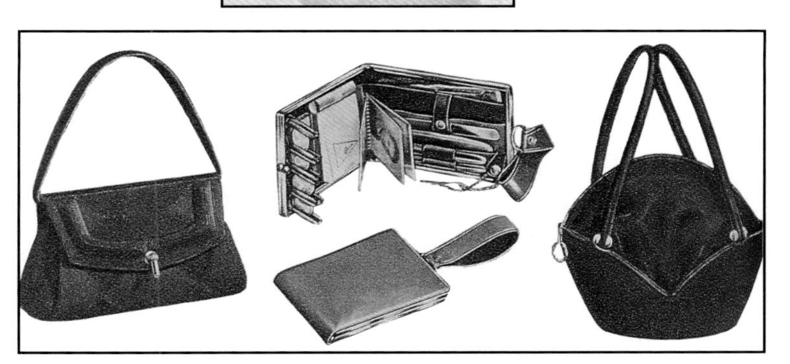

Corde Vagabond. Luxurious genuine corde. 8.5 x 4.75 inches. $6 [$85-90] **Forget-me-not Carryall.** Complete with ball point pen, mechanical pencil, billfold pocket, celluloid picture holder, coin holder, memo pad, removable key chain. In calf-like plastic. 4 x 6.5 inches. $2.28 [$25-30] **The Glamourette.** Look what's happened to the pouch. Cleverly rounded, shirred into a flare-out bottom cuff. In rayon velvet. 9 x 8 inches. $3.97 [$55-60]

Sueded Rayon Bag 'n' Gloves. Shirred vanity-pouch, gold-color metal frame. 9.5 x 4.75 inches. Gloves, 10 inches long. $4.99 set [$40-45 set] **Roomy Diplomat.** Smart satchel in calf-like plastic. 12.5 x 10 inches. $3.59 [$35-40] **Scalloped Corde Success.** 10.25 x 6.25 inches. $6.79 [$85-95]

Spring/Summer 1952 *Fall/Winter 1952*

Note: *Some of these styles had been offered by Sears as early as 1950, such as the "Pouch Star" which was originally faille and is now made of rayon velvet. Cloth, faille, and rayon velvet in these "bargain priced handbags" tend to bring higher prices than the "calf-like plastic," which sometimes discolors, dries, and cracks. Many of these plastic pocketbooks are, however, found in excellent condition, which may reflect the durability of plastic!*

Tapestry Box Bag. Rich colors blend in a beautiful cotton embroidered top. Eight-sided rayon faille box bag, full-view mirror inside top. 5.5 x 4 inches. Rayon lined. $3.59 [$65-70]

"Peek-a-boo" Box Bag. Clear Lucite with graceful leaf design. Lavishly pleated. 7.5 x 4 inches. $3.79 [$85-90]

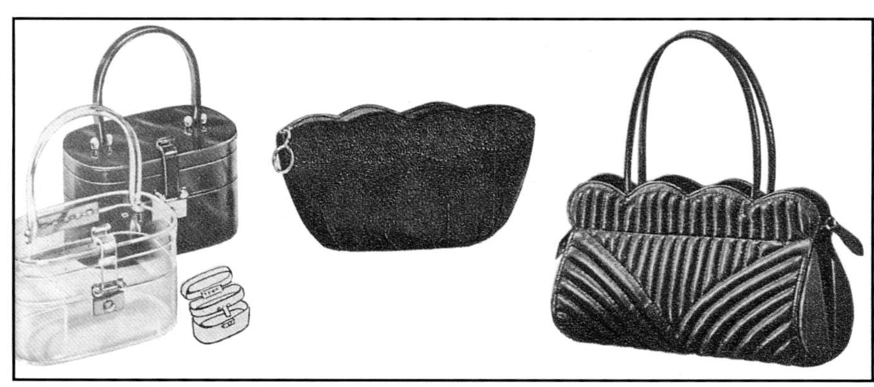

Double Decker Box Bag. Glamour show-off made of styron molded plastic—a modern miracle. Glass-clear or amber-colored. 7.5 x 4 inches. $3.59 [$65-85] **Corde Clutch Bag.** Fetchingly feminine. Smartly scalloped. 9.25 x 5.5 inches. $3.59 [$65-70] **"Quilted" Leather Bag.** Quilted (trapunto) effect turns this scalloped zip-top bag into a dressy accent. 10 x 6.25 inches. $5.50 [$65-70]

Bag and Belt. Casual saddle-stitched shoulder-strap bag in calf-like plastic with metal crest. 9.25 x 6.25 inches. Matching fob belt, 2.25 inches wide. $2.49 [$40-45] **Leather Vagabond-Pouch.** Eye-catching V flap-over tab. 8.75 x 4.75 inches. $3.79 [$45-50] **Superb Shoulder-strap Bag.** Copy of an exclusive English bag. Calf-like plastic. 9.75 x 9.25 inches. $3.59 [$65-70]

Pleasing Pouch. Softly shirred leather. 9.75 x 6.5 inches. $3.59 [$40-45] **Drawstring Shoulder-strap Bag** in genuine leather. 9.5 x 8.75 inches. $3.97 [$50-55] **Curving Box Bag.** Full-view mirror. Genuine leather. 9 x 5 inches. $3.79 [$45-50] **Box Bag.** Soft leather with side pleats and scalloped flap-over top. 7 x 4 inches. $2.69 [$45-50] **Shoulder-strap Bag.** Leather. 8.75 x 6.5 inches. $4.99 [$35-40] **Drawstring Pouch Favorite.** Softly gathered leather with matching plastic handles. 8.75 x 8.5 inches. $2.99 [$40-45] **New-look Vanity. Leather.** 7 x 5.5 inches. $3.79 [$40-45] **Scalloped Top Handler.** Leather. 10.5 x 7 inches. $3.97 [$40-45]

***Left:* Swinging Shoulder Bag.** In smooth calf-like plastic. Rayon lined. 10 x 7.25 inches. $2.69 ***Top row:* Pouch Star** in lush rayon velvet, bright-touched with sparkling Lucite accents. 9 x 5.75 inches. $2.38 [$40-45] **Delightfully Different Pouch.** Calf-like plastic. Flare-out top slims to a straight bottom. Center attraction is like a flower petal. Bracelet handles. 10.5 x 7 inches. $2.38 [$25-30] **Box Bag** in calf-like plastic. 8.75 x 4.75 inches. $1.99 [$25-30] ***Center row:* The Vain Vanity** with stand-up collar, bottom tier. Suede-like plastic. 7.75 x 6 inches. $2.69 [$25-30] **Petite Clutch Bag** in dressy rayon velvet. 8.5 x 5 inches. $1.27 [$30-35] **Hold-everything Satchel.** Calf-like plastic. 8.75 x 6.25 inches. $2.69 [$20-25] ***Bottom row:* The Satchel** that's a smash-hit. Saddle-stitching. In Texon, a coated fibrous material that looks like saddle leather. 8.5 x 6 inches. $2.99 [$18-20] **Ruffled Pouch.** Stand-up collar and smart dressmaker touches. In calf-like plastic. 10.5 x 7.25 inches. $2.38 [$20-25] **Dream of a Drawstring Pouch** in suede-like plastic. 10 x 8.5 inches. $1.99 [$20-25]

***Top row:* Surprise Pouch.** Calf-like plastic. 12 x 6 inches. $1.99 [$20-25] **Classic Frame Bag.** Bracelet handles. Calf-like plastic. 12 x 7.25 inches. $2.69 [$20-25] **Terrific Vanity.** Double corded handles. Calf-like plastic. 7.25 x 7 inches. $2.99 [$20-25] ***Center row:* Popular Box Bag** in calf-like plastic. 9 x 4.5 inches. $2.69 [$25-30] So in Love with the Satchel. Pretty stand-up collar, appliquéd crescents, and anchor top handles. Calf-like plastic. 9.5 x 5.5 inches. $1.99 [$20-25] **Medallion Shoulder-strap Bag.** Calf-like plastic. 8.25 x 8.5 inches. $2.69 [$20-25] ***Bottom row:* Stupendous Shoulder-strap Bag** in calf-like plastic. 10.5 x 8 inches. $2.99 [$18-20] **Perky Pouch** wears a pretty stand-up collar. 8 x 5.5 inches. In cotton and rayon guimpe $1.99 [$40-45] In rayon pleated faille. $1.99 [$40-45] **Zip-top Delight.** Spacious underarm bag in calf-like plastic. 13 x 6.5 inches. $1.99 [$20-25]

Note: Plastic handbags for summer use are more interesting to collectors than the same handbags in dark colors for winter. Although the plastic may fade or discolor in time, these pocketbooks can be cleaned with soap and water for a fresher look.

Top row: **Washable Slip-cover Bag.** Button centered daisies. 9.5 x 5.5 inches. $1.99 [$28-32] **Leaf-lovely Pouch.** Plastic with embroidered linen look. 8.75 x 4 inches. $1.99 [$28-3] ***Center row:*** **Surprise Pouch.** Calf-like plastic. 10.5-6.25 inches. $1.99 [$25-30] **Scalloped Success.** Calf-like plastic. 8 x 3.75 inches. $1.99 [35-40] **Bottom row: Three-dimensional** satchel. Stand-out leaf pattern on plastic. 8 x 4.75 inches. $1.99 [$35-40] **Adjust-able Shoulder-strap Bag.** Calf-like plastic. 8.25 x 7 inches. $1.99 [$18-20] **Drawstring Darling. Bracelet** handles. Calf-like plastic. 9.75 x 8.5 inches. $1.99 [$20-25]

A: Wondrous Plastic Vanity. Made of hardy plastic cubes. 7 x 4.5 inches. $3.79 [$40-45] ***B:*** **Nylon News in a Drawstring.** 100 percent nylon. 8 x 9 inches. $2.99 [$18-20] **C: Bamboo Box Bag.** Sensible wipe-clean plastic. 8 x 4.75 inches. $2.69 [$45-50] ***D:*** **Pique Drawstring.** Eyelet ruffled cotton pique bag. 11 x 9.5 inches. $2.39 [$30-35] ***E:*** **Hand-crocheted Toyo Straw Pouch.** Straw-like lining. Made in Japan. 12 x 8.5 inches. $3.59 [$40-45] ***F:*** **Plastic Coil Favorite.** 9 x 5.25 inches. $3.59 [$40-45]

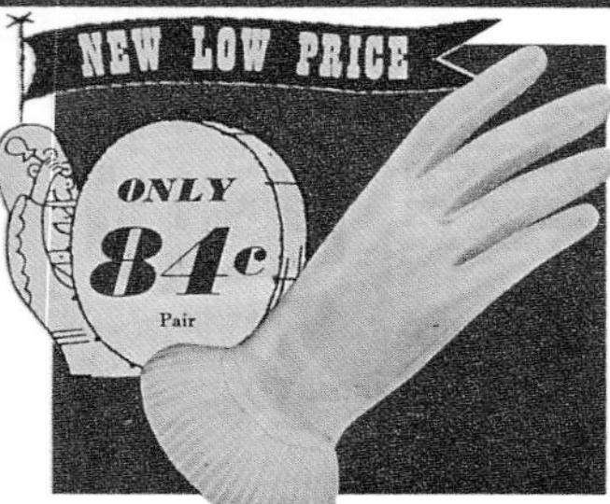

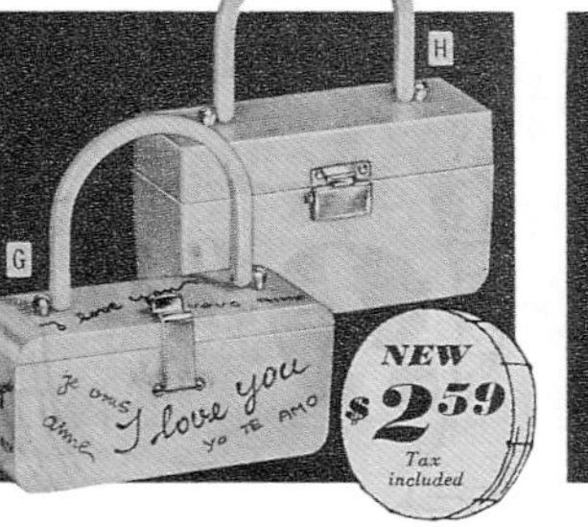

Nylon Slip-on. A precious nylon glove at a rock bottom price. Length 9.75 inches. $.84. [$8-10] **Button-up Bag.** Multiplies to three. Linen-like slip-cover is navy blue on one side (acetate and rayon). Just unbutton and reverse to show sparkling white side (100 percent cotton), or carry white plastic bag without slip-cover for still another accent. 10 x 7 inches. $1.99 [$20-25] **Waist Cincher with Winged Pouch.** Cincher whittles inches off your middle. In floral design or all white. Plastic belt is 3 inches wide; bag is 8 x 5.5 inches. $2.49 [$30-35]

Scalloped Success in leather-like plastic. 10.5 x 6.5 inches. $2.69 [$20-25] **Summer Romance** with the look of mother-of-pearl sparkle. Two versions: "I love you" scribed in French and many other languages and the other, plain-as-a-pearl. Both of Styron molded plastic. $2.59 [$65-70] **Good Traveler.** Sturdy denim or summery bamboo that's really made of plastic. 12.25 x 6 inches. $3.59 [$45-50]

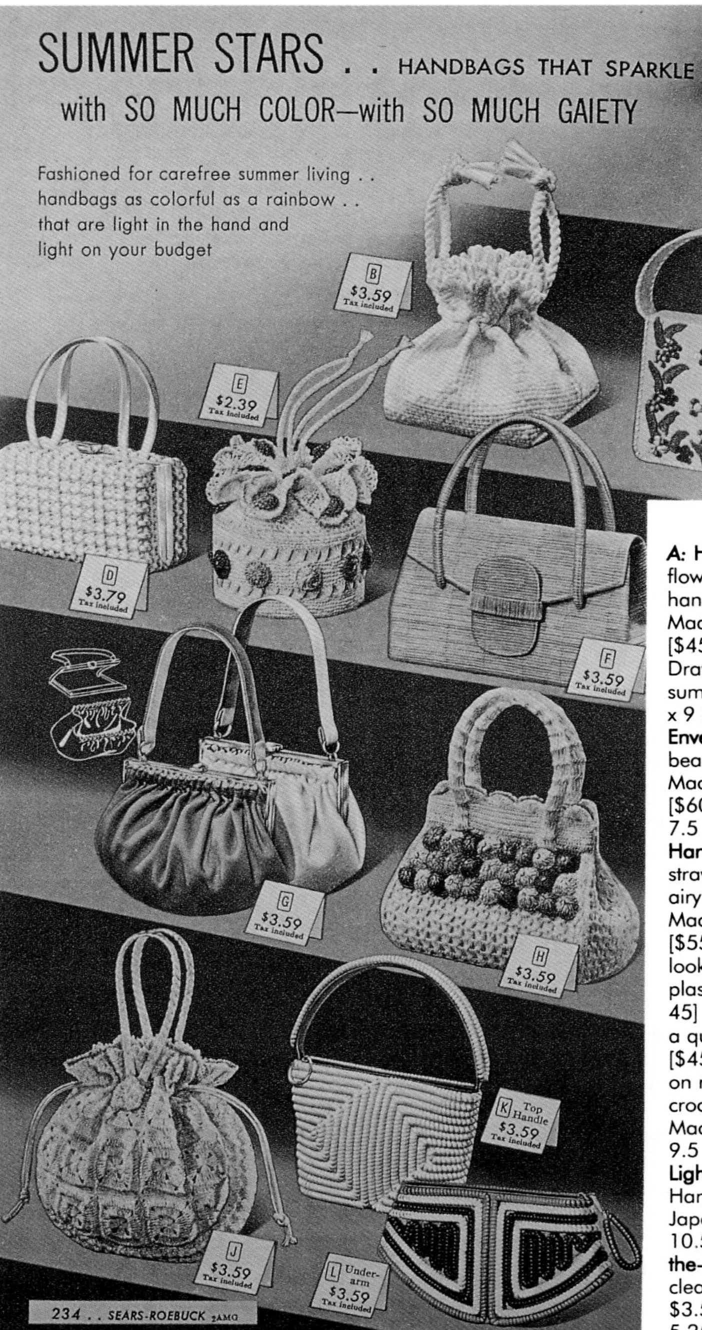

A: Happy Straw Box Bag. Blooming flowers, real as nature's own, are hand embroidered on toyo straw. Made in Japan. 8 x 5 inches. $2.39 [$45-50] ***B:* Washable Nylon Nifty.** Drawstring bag that's a carefree summery accent. 100 percent nylon. 8 x 9 inches. $3.59 [$30-35] ***C:* Beaded Envelope.** Rainbow-colored wooden beads on front of toyo straw bag. Made in Japan. 11 x 6 inches. $3.59 [$60-65] ***D:* Wondrous Plastic Vanity.** 7.5 x 4 inches. $3.79 [$45-50] ***E:* Hand-crocheted Drawstring.** Toyo straw, hand crocheted for the look of airy lace, lavished with dainty rosettes. Made in Japan. 6 x 6 inches. $2.39 [$55-60] ***F:* Sensible Satchel** with the look, the feel of bamboo, but it's plastic. 11 x 6.5 inches. $3.59 [$40-45] ***G:* Reversible Curtain Rod Bag.** For a quick change, reverse cover. $3.59 [$45-50] ***H:* Stepping-out Pouch.** Row on row of rosettes on toyo straw, hand crocheted for a light hearted lacy look. Made in Japan. Rayon lined. About 9.5 x 7.25 inches. $3.59 [$45-50] ***J:* Light-hearted Drawstring Pouch.** Hand-crocheted toyo straw. Made in Japan. Straw-like lining. About 14 x 10.5 inches. $3.59 [$45-50] ***K:* On-the-go Zip Tops** in plastic coils that stay clean. Top Handle. 9 x 5.75 inches. $3.59 [$40-45] ***L:* Underarm.** 10 x 5.25 inches. $3.59 [$40-45]

***Clockwise from top:* Salt Box Bag.** Leather with full-view mirror. 6.5 x 5.25 inches. $3.59 [$50-55] ***B:* Go-everywhere Pouch.** Softly shirred leather, practical three-part frame. 11 x 5.5 inches. $3.59 [$40-45] **A: Shirred Box Bag.** Leather. 9.5 x 4.25 inches. $3.59 [$50-55] **Shoulder-strap Drawstring.** Soft leather, plastic draw-string. 9.25 x 9.25 inches. $3.79 [$45-50]

Note: Sears offered an amazing diversity of styles in handbags during a very few years. We find an incredible range of box style bags, for example the "shutterbox bag" and "shirred box bag." Women carried pocketbooks to match each change of shoes and wore hats to accessorize both.

Clockwise from top: "Intrigue." An exact copy of a French design. Leather with gold-color metal lock. 11.25 x 5.25 inches. $4.99 [$40-45] **Successful Shoulder-strap Bag.** Leather. $4.49 [$40-45] **Good-Shape Top Handler.** Leather. 14.5 x 7 inches. $3.79 [$40-45] **Newsy Vanity.** Leather. 7 x 5.5 inches. $3.59 [$50-55]

Top row: Screened Vanity. Peek-a-boo rayon mesh screens front of a calf-like plastic vanity. 5 x 7 inches. $2.69 [$25-30] **Adjustable Shoulder-strap Bag** flaunts laced flap-over top. Calf-like plastic. 9 x 6 inches. $2.39 [$20-25] **Box Bag.** Rayon faille. Full-view mirror. 8.25 x 3.5 inches. $2.39 [$40-45] ***Center row:*Shutter Box Bag,** the most unusual of the season. Shutter-like side panels contrast with smooth calf-like plastic. Full-view mirror. 7.5 x 4.25 inches. $2.69 [$30-35] **The Collared Pouch** in two fabrics. 8.5 x 6 inches. Easy cleaning plastic. $1.99 [$20-25] **Dress-up Rayon Faille.** $1.99 [$40-45] **Fashion Plate Pouch** with gold-color metal accents. Calf-like plastic. 11.25 x 5.25 inches. $2.99 [$20-25] ***Bottom row*: Shirred Pouch** in calf-like plastic. 10 x 6.5 inches. $1.99 [$25-30] **Drawstring Delight.** Suede-like plastic. 9.5 x 9 inches. $1.99 [$25-30] **Multiple-frame Bag** with pockets galore. Calf-like plastic. About 11.5 x 7 inches. $3.29 [$20-22]

Circulate in a Glittering Set. Rayon velvet pouch, dotted with multi-colored sequins on the front and a sueded rayon glove. $4.99 [$60-65] **Fashionable Trio.** Ruffled and pleated sueded rayon pouch and ballerina cuffed sueded rayon glove, plus a gold-color metal glove ring to keep an eye on your gloves. $3.59 [$50-55] **Barrel of Fashion.** A barrel-shaped vanity, pleated in sueded rayon, with matching convertible cuffed glove. $3.99 [$45-50]

Key to a Shoulder Bag. Adjustable shoulder-strap bag in calf-like plastic. Security lock with shining key. 11 x 6.75 inches. $3.59 [$20-25] **Stepping-out Shoulder Bag.** Adjustable shoulder strap. Calfskin. 9.25 x 5.5 inches. $8 [$40-45]

Flirtatious Faille Box Bag. Pleated acetate and rayon faille. Leather lined. Full-view mirror. 8 x 4 inches. $3.59 [$50-55] **"Zippy" Clutch Bag.** Rayon faille. Attached coin purse, mirror. 11.25 x 5.25 inches. $2.69 [$35-40]

"Brief Case" Shoulder Bag. All-around zipper underneath flap-over top reveals compartment with pencil, mirror, memo pad, comb, three pockets. Calf-like plastic. Adjustable shoulder strap. 8.5 x 7 inches. $3.59 [$20-25] **Superb Shoulder Bag.** An exact copy of an English original. Calf-like plastic. 10.25 x 7.25 inches. $3.79 [$20-25]

Filigree Jewel Box Bag. Acetate and rayon faille with gold-color metal filigree. Full-view mirror. 7.75 x 4 inches. $3.59 [$60-65] **Toting Calfskin Satchel.** Leather lining. 10 x 4 inches. $8.50 [$40-45]

Beauteous Barrel Bag. Genuine leather. 8.5 x 4.25 inches. $3.79 [$45-50] **Faille Kangaroo Pouch.** Acetate and rayon faille, every inch lined with genuine leather. Mirror. 8.52 x 6.5 inches. $3.59 [$40-45]

Campus Favorite in Cowhide. Adjustable shoulder strap. 9.5 x 5.75 inches. $3.97 [$35-40] **Fabulous is the Word.** Inspired by a fabulous French original. Leather-like plastic. Attached coin purse. 11 x 5.25 inches. $3.79 [$20-25]

Stripe-happy Scarf. Acetate and rayon taffeta. Bold accent for the woman who loves color; to call attention to suits, coats, and dresses. $1.69 [$25-28]

French Fireworks. Dazzling rhinestone stars on rich rayon velvet ascot, as gay as Paris. Rayon taffeta lining. About 33.5 inches long. $1.89 [$25-28]

Ribbon Flair. Slip-through neckline novelty. Acetate and rayon ribbon from France, about 5.5 x 35 inches. $.98 [$20-24]

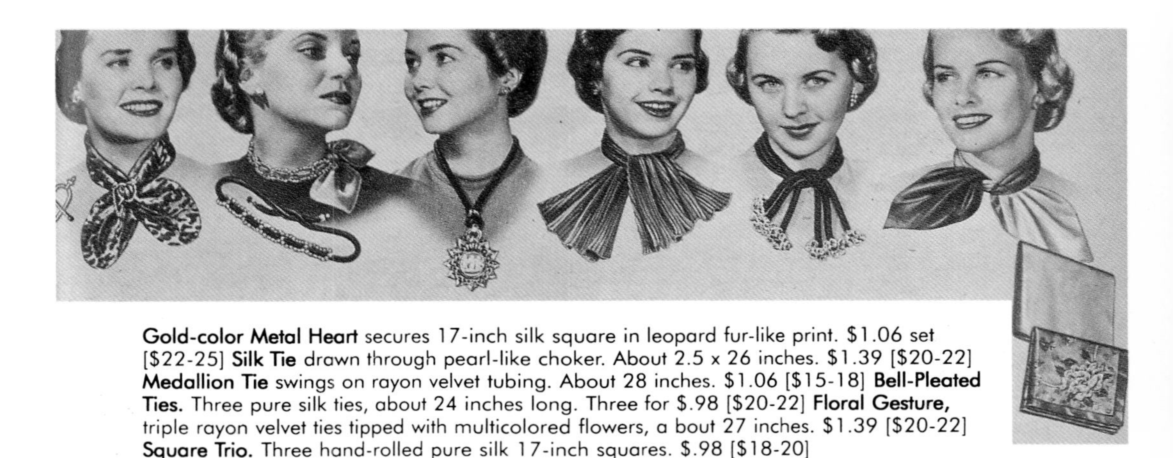

Gold-color Metal Heart secures 17-inch silk square in leopard fur-like print. $1.06 set [$22-25] **Silk Tie** drawn through pearl-like choker. About 2.5 x 26 inches. $1.39 [$20-22] **Medallion Tie** swings on rayon velvet tubing. About 28 inches. $1.06 [$15-18] **Bell-Pleated Ties.** Three pure silk ties, about 24 inches long. Three for $.98 [$20-22] **Floral Gesture,** triple rayon velvet ties tipped with multicolored flowers, a bout 27 inches. $1.39 [$20-22] **Square Trio.** Three hand-rolled pure silk 17-inch squares. $.98 [$18-20]

Note: The early 1950s look combined small silk scarves and other neck pieces with cashmere or orlon short-sleeved sweaters. Some girls wore a complete sweater set with a button-up matching sweater over a short-sleeved pull-over. Belts were popular, especially the wide leather "contour" styles and the "cincher."

Colorful Flowers hand-painted on silk. Three sizes: 17 and 32 inch squares and the long 17 x 44 inch scarf. Long Scarf: $1 [$20-25] 17-inch with rhinestone accent: $.88 [$15-18] Large square: $1 [$20-25]

Duet in Plaid, 55 percent wool, 45 percent rayon. Hug-you-tight helmet is rayon lined. Cotton fringed scarf is 6.5 x 39 inches. $2.49 set [$15-18 set]

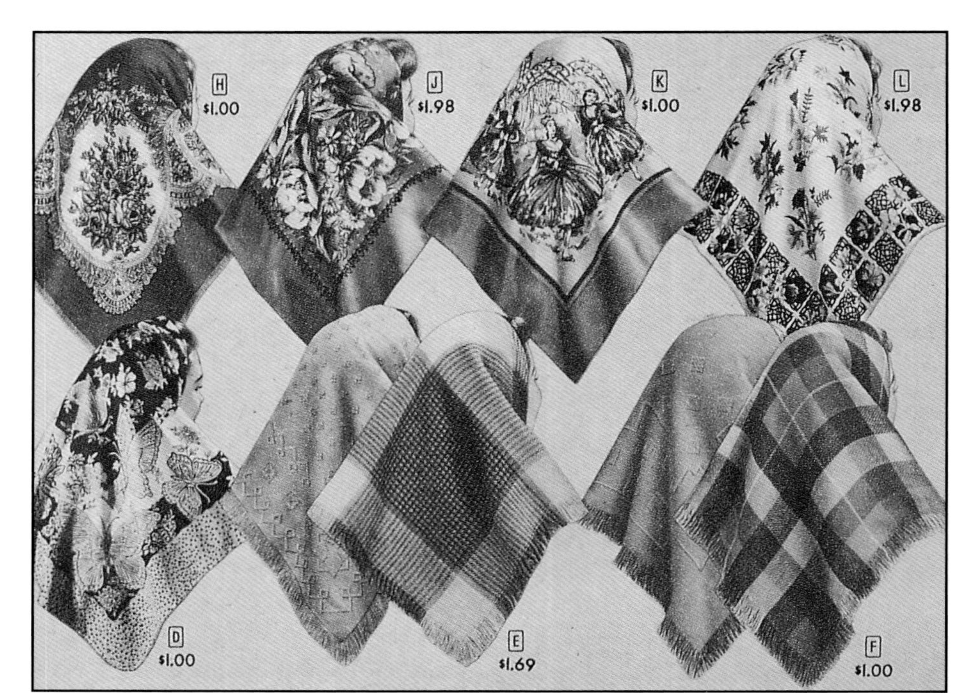

Paisley and Floral Combine. Spun rayon square. $1 [$15-18] **A Silk Bouquet.** Silk crepe, each corner different, from Japan, 35 inches square. $1.98 [$18-20] **Dancing Ballerinas.** Hand-rolled silk, 35 inches square, from Japan. $1 [$20-22] **Tone-on-tone** silk, 32 inches square. $1.98 [$20-22] **Wool Head Warmer.** 30-inch square. $1 [$8-10] **Zephyr Wool Square.** 33 inches square. $1.69 [$8-10] **Golden Touch Square.** Spun rayon, 35 inches square. $1 [$20-22]

Soft Angora Touch. Silk square from Japan. $1 [$20-22] **Velvet and Silk.** Sham pearl studded black rayon velvet loop-the-loop tubing add a fanfare of pleated silk ties. $1 [$20-22] **Velvet Tie.** Plastic shells, sparklers on rayon velvet tie. $1 [$18-20] **Wardrobe of Three** hand-rolled silk squares, from Japan. $1 [$18-20] **Medallion Rage.** On black rayon boucle cord. $1 [$15-18] **Artistic Triple Tie.** Three silk ties. $1 [$20-22]

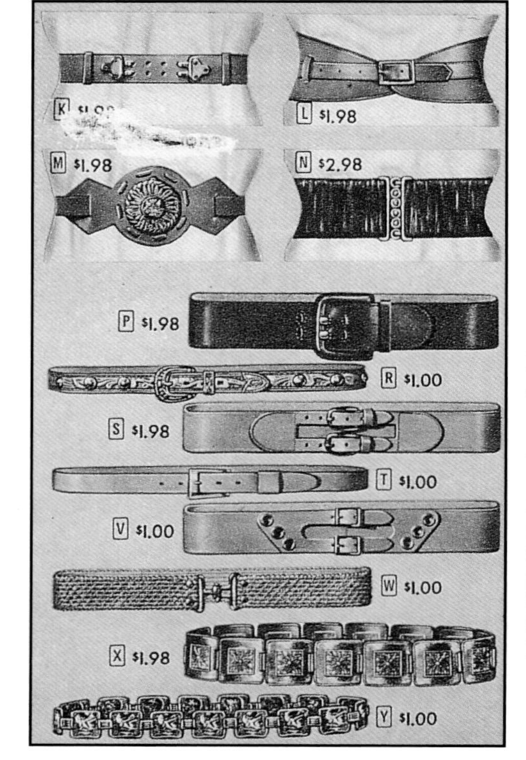

Soft Cowhide for wee waists. About 2.25 inches wide. $1.98 [$16-18] Embossed Cowhide. $1 [$18-20] **Cowhide Polo.** Wide and handsome. $1.98 [$18-22] **Top-grain Cowhide Belt.** $1 [$12-15] Neolite Polo. $1 [$18-22] **Glitter Waist Cincher.** $1 [$15-18] **Lightweight Belt Champion.** Aluminum. $1.98 [$18-20] **Heraldic Disc Belt.** $1 [$15-18]

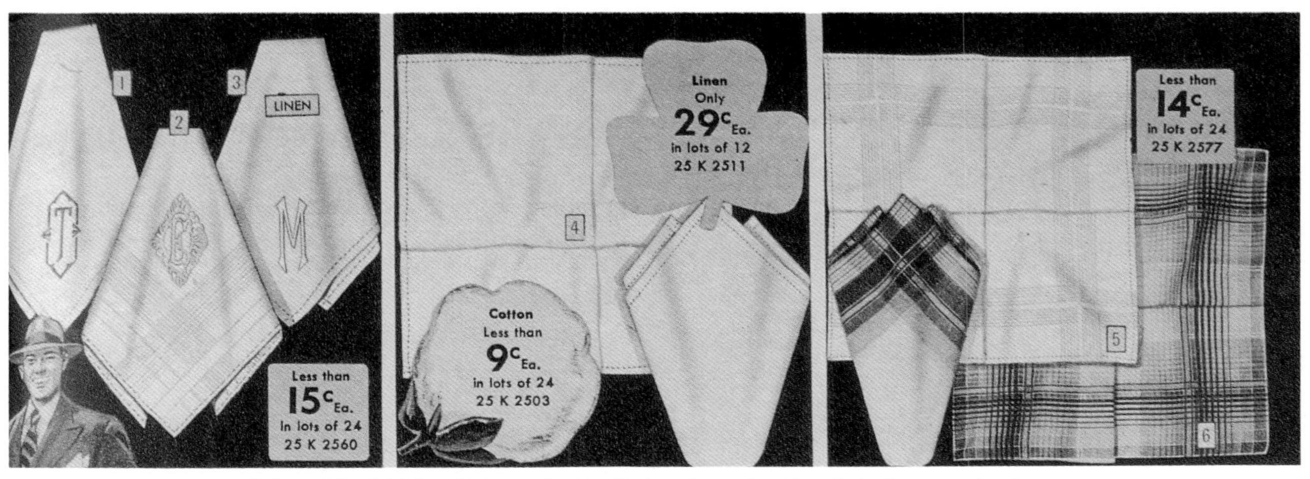

***Left to right*: Initialed.** Make sorting family laundry easier. **Hemstitched** on good white cotton. 12 for $1.95, 24 for $3.40 [$10-12 each] Genuine Irish Linen. 6 for $2.40, 12 for $4.30 [$12-15 each] **Woven Borders.** White and colored. 24 for $2-3.39 [$10-12]

Note: The early 1950s was the pre-Kleenex era. Men and boys carried handkerchiefs, women and girls tucked handkerchiefs into their pockets and handbags. It's hard to imagine all that washing and ironing. In my house, little stacks of handkerchiefs were ironed for everyone in the family and it was the first thing I learned to iron as a little girl.

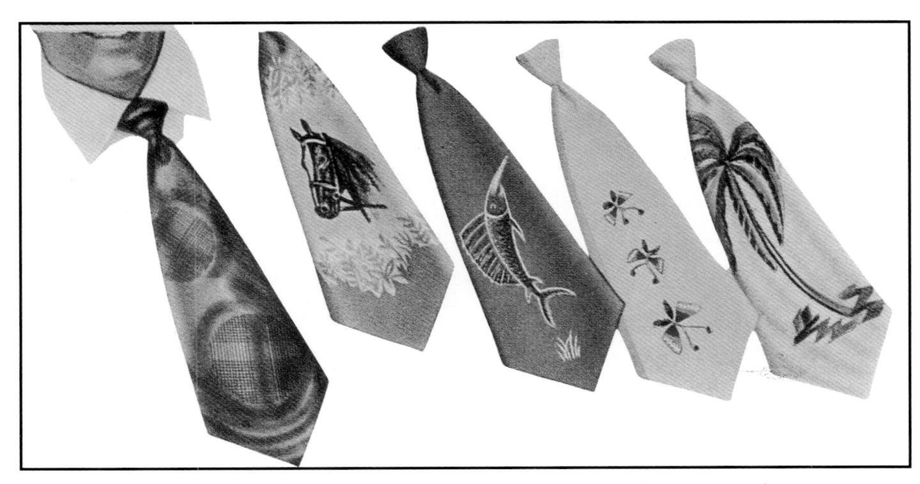

Hand-painted Designs. Rayon, wool lined $.94-1.44 [$15-30] **Wool, Rayon Foulards.** All are wool lined—to tie well shaped knots, to resist wrinkling. 48 inches long. Wool ties and tie are 4.5 inches wide.

Wool Worsted Ties in solid colors. $.94 [$15-30]

Note: There is a market for vintage 1950s ties for men. The more unusual styles, such as the palm tree design or the geometric, tend to bring the highest prices. Men who enjoy less-traditional neckwear sometimes collect vintage ties. Women enjoy the ties; sometimes wearing them as belts.

Rayon and Silks. $.97-1.97 [$15-30]

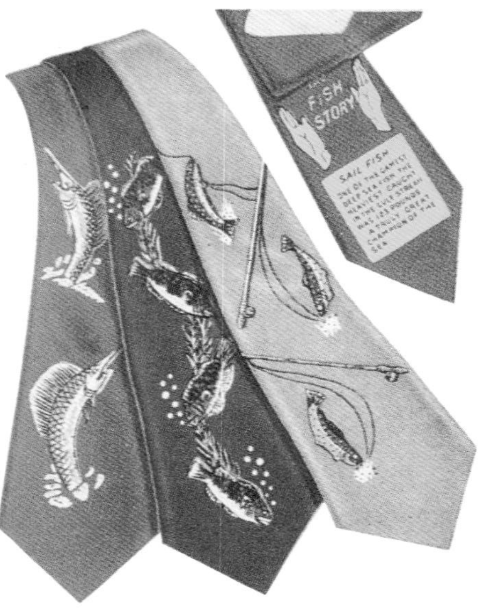

Game Fish. $1.47 each [$15-30]

Hand Painted. $.97-1.97 [$15-30]

Game Birds. $.97 each [$15-30]

Hand-painted Ties. Acetate, wool lined. $.97. [$50-65]

Our Finest Acetates. Hand painted and screen-printed. $1.47 [$15-65]

Pure Silks. $1.97 [$15-30]

Good. Rich acetate, dobby, jacquard, and satin fabrics. Wool lined, white acetate-tipped ends. $1 [$15-30]

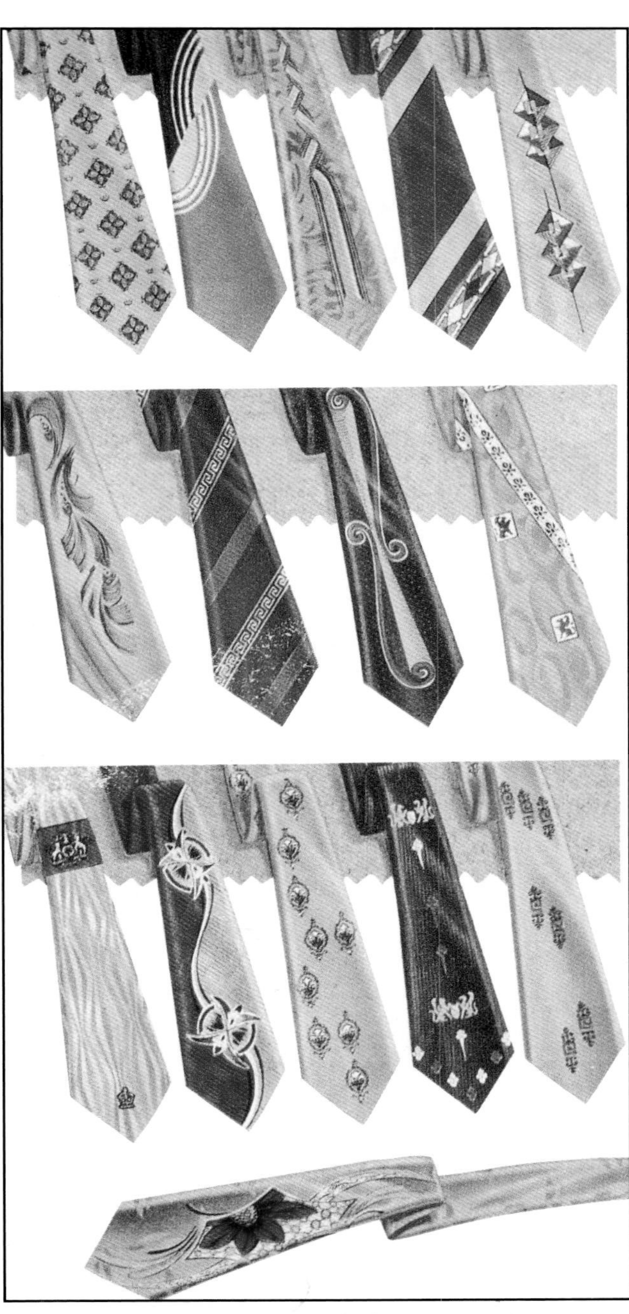

Better. Luxurious Acetate, foulards, jacquards, twills, oxford cloth, and crepe. Wool lined, white acetate-tipped ends. $1.50 [$15-30]

Our Best. Pure silk crepes and jacquards, wool lined, acetate-tipped ends. $2.50 [$30-60]

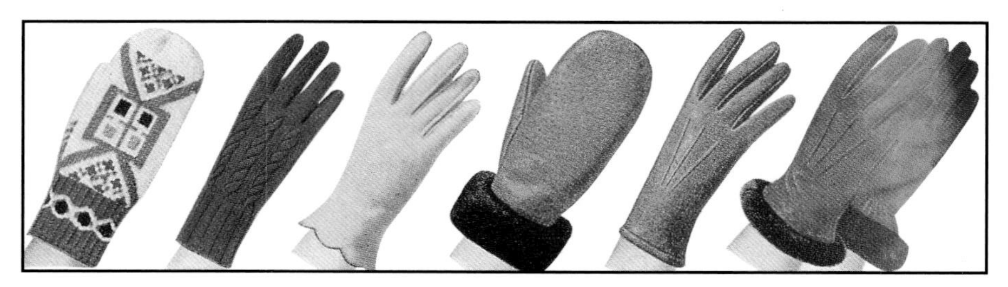

Gloves, Mittens. Warm woolens, lined leathers, and a dressy fabric glove. $.98-2.98 [$4-12]

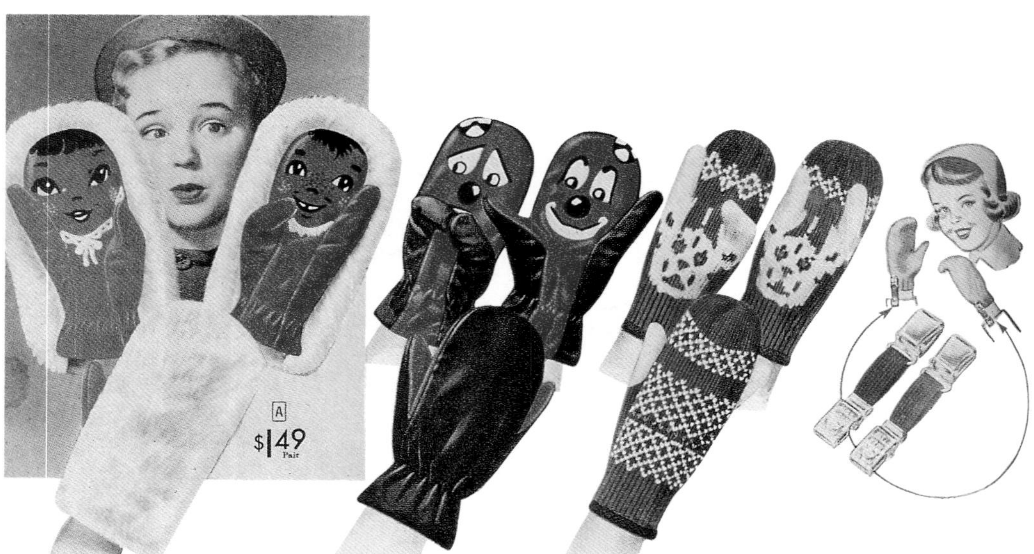

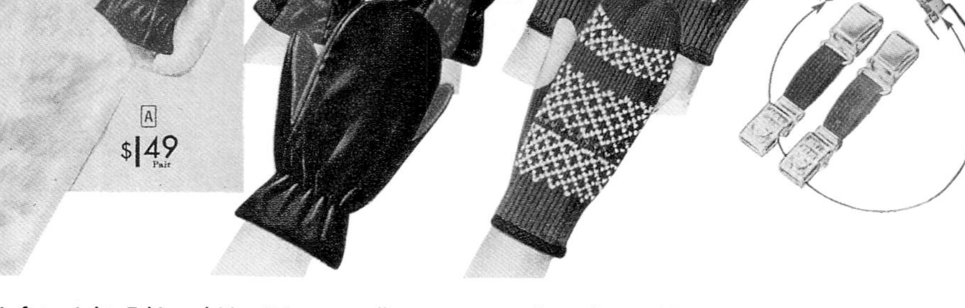

Left to right: **Eski and Mo.** Water-repellent cotton poplin palms, white bunny fur backs. $1.49 [$40-45] **Teary and Cheery** and Floppy the Puppy. 100 percent wool jacquard. $.98 pair [$20-25] **Mit Clips.** $.44 [$4-5]

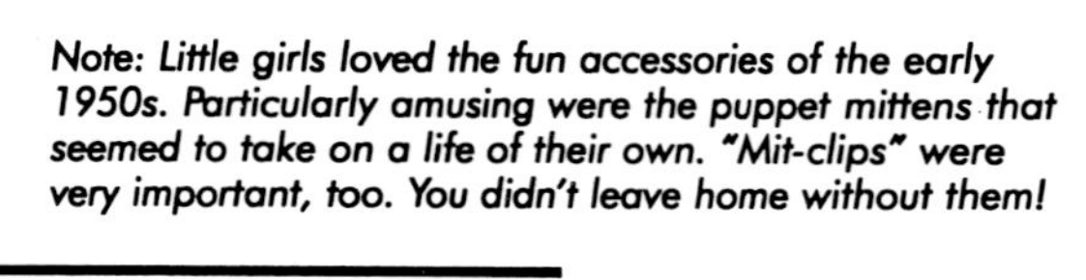

Note: Little girls loved the fun accessories of the early 1950s. Particularly amusing were the puppet mittens that seemed to take on a life of their own. "Mit-clips" were very important, too. You didn't leave home without them!

Handbag Wardrobes. $1.28-5.50 [$15-25]

Sears Head Toppers. For the warmth she needs, for the color she wants. $.49-98 [$8-30]

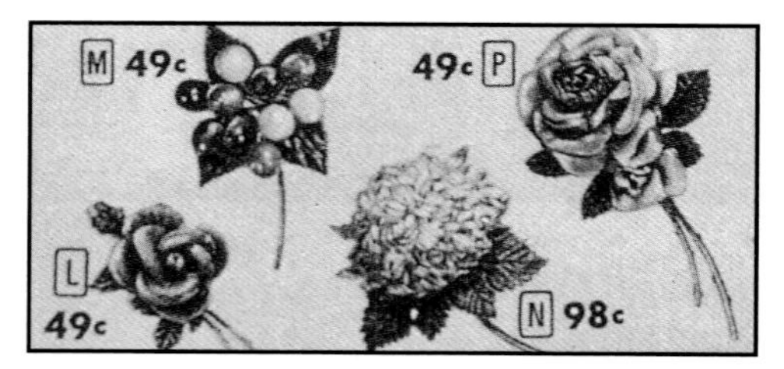

Fashion-fresh Accents, the easy, inexpensive way to perk up your wardrobe. Priced $.49-.98 [$5-20]

***K:* Enchanting Scoop Brim Bonnet.** Gleaming straw-effect cellophane, rich rayon velvet. $4.98 [$20-22] ***L:* Beautifully Woven Rustic Straw Bonnet** with pastel flowers. $5.98 [$24-28] ***M:* Pretty Profile Beret** with flowers, rayon taffeta loops, straw braid. $3.98 [$20-22] ***N:* Poppies on Big, Beautiful Picture Brim.** Open crowned sisal straw, edged with sheer braid. $5.98 [$40-45] ***P:* Profile Brim.** Feather-flower fancy. Fine straw braid. $4.59 [$35-40]

***R:* Exquisite Flower Halo.** Open crown. Straw-like cloth. $4.98 [$30-35] ***S:* Winged Sailor.** Sheer nylon tulle puff, band around crown. Straw braid. $2.98 [$30-35] ***T:* Rich Flower/Feather Fancy,** fine straw braid. $3.98 [$20-24] ***V:* Flattering Off-the-face Brim.** Straw braid. $4.49 [$30-35]

Sparkling White Polka Dots. Detachable scarf buttons on wedding ring calot. Rayon crepe. $2.59 [$20-25]

Note: Millinery from the early 1950s doesn't generally bring high prices in the vintage clothing market. The highest prices are paid for designer hats, wide-brimmed styles, or hats that have spectacular trim, especially feathers. Cocktail hats are beginning to catch on with collectors who wear them for occasions such as weddings, art exhibits, and cocktail parties.

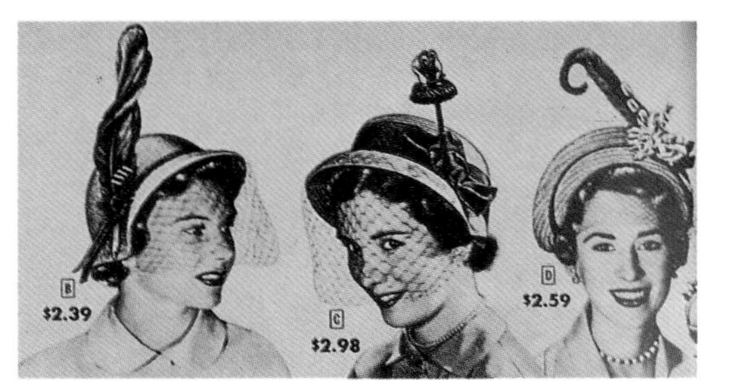

Spiraled Quills, crisp grosgrain band on smart straw braid cloche. $2.39 [$20-22] **Straw Braid Cloche** has dainty flower atop grosgrain ribbon stick-up. $2.98 [$20-22] **Flattering Profile Beret.** Straw braid with pretty feather/flower fancy. $2.59 [$25-28]

Flowered Bonnet half-hat, ribbon ties under your chin; rayon veiling over buckram crown. $2.49 [$20-22] **Glamour Calot.** Straw braid stitched in leaf design; dotted rayon veil. $1.94 [$35-40] **Head-hugging** tiered helmet. Big side bows and band of rayon taffeta. Straw braid. $2.39 [$30-35] **Sparkling White** pique butterfly bow and facing across off-the-face brim. Shiny straw braid. $3.98 [$30-35] **Visor Cap** of cool straw-like braid. $1.98 [$20-22]

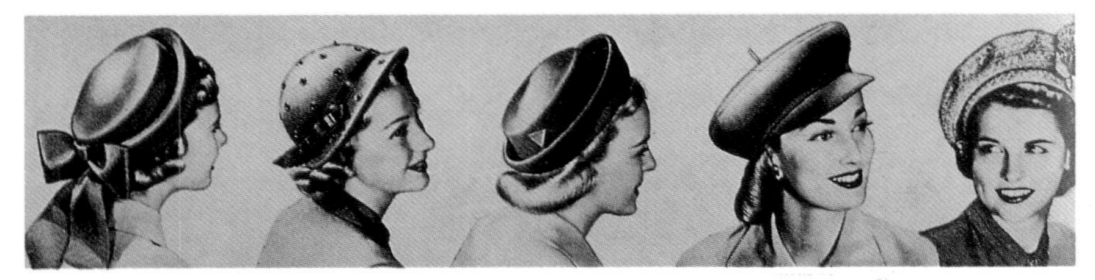

100 Percent Wool Felt Bumper, back bow and streamers of wide grosgrain ribbon. $1.98 [$20-22] **Eyelet Cloche.** 100 percent wool felt, grosgrain ribbon band, bow at side. $2.49 [$20-22] **100 Percent Wool Felt.** Popular Breton with smart lines, matching hat pin. $1.74 [$20-22] **Bloused Crown Visor Cap** of 70 percent wool, 30 percent cotton felt. Self stick-up on crown. $1 [$20-22] **Pretty Dimpled Beret** of straw-like braid, saucy stick-up at side. $2.19 [$22-24]

Wedding Ring Cloche with fashion-rich rayon velvet touches, draped band, dashing wings. $3.98 [$30-35] **Profile-Pretty Beret,** draped for flattery. $1.98 [$25-30] **Gold-touched Hamburg.** Wide grosgrain band. Rayon veil. $2.98 [$20-25]

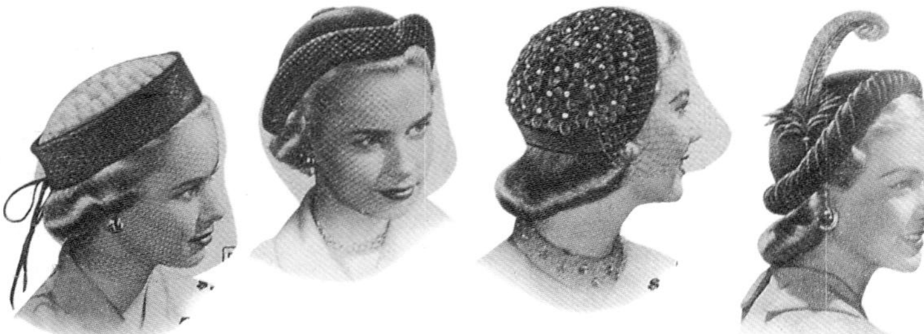

Dimpled Brim Bonnet in rayon velvet adorned with rayon veil, chenille dotted border. $2.98 [$25-30] **Feathered Pill Box,** rayon velvet-cuff. Rayon cage veil. $3.39 [$18-20] **Tricorne** triumphs in rich rayon velvet, starring the enchanting sweetheart cuff. Crisp rayon veil. $3.29 [$20-25] **Flowered Helmet** in rich rayon velvet and veil. $3.49 [$20-25] **Draped Wedding Ring** in rayon velvet with curving feather. $2.98 [$20-25]

Poke Bonnet in wool felt with rayon velvet covered buttons. Chenille dotted rayon veil. $2.98 [$25-30] **Glitter Calot** in wool felt. $1.69 [$35-40] **Pearl-of-a-Bonnet.** Pearl-like beads and rayon velvet buttons on wool felt. Rayon veil. $3.98 [$25-30]

Note: Highest prices are paid for 1950s hats in excellent condition, with feather trim. Veils add dramatic flair, and rhinestone detailing is a plus. Women do not buy hats that they perceive would make them look silly Only the most confident women wear fashion millinery. Beyond the obvious weather protection, fashion hats tend to attract attention to the wearer. Dealers who buy fashion hats for resale as wearables should keep image in mind. How would this hat look? Who would wear it? Do I have potential customers for this? Some collectors buy hats for decoration, to put on a hat rack or hat stand in their homes. Costumers stock vintage millinery to use in projects, such as film, television commercials, or (print) advertisements.

Note: Young girls' hats of the early 1950s were trimmed conservatively and were available in a number of flattering styles. The market is not strong for these hats, unless there is a complete set, with a matching coat. Head sizes are considerably smaller than most women collectors wear.

Shell Bonnet, a new variation of the popular cover-up. Stitched rayon suede cloth. $2.39 [$20-22] **"Starlet" Bunny Fur Set.** Includes clip-on halo band, pull-through scarf, and warm mittens (cotton fleece-lining) with red rayon palms. $4.98 [$30-35]

Cotton Velveteen Brim on wool felt cloche. Fluffy, natural feather trim. $2.98 [$18-20] **Perky Feather Stick-up** on wool felt hat, cuffed for interest. $2.98 [$18-20] **Popular Peak Cap,** just like big sister's. Pinwale corduroy with three self-covered buttons. $2.19 [$18-20] **Look Picture Pretty** in our wool felt flatter, sporting a feather fancy and self band. $2.98 [$18-20] **Cute Chukker** in wool felt with pheasant feather. Self band, bow. $2.98 [$18-20]

Dutch Cap in favored corduroy. $2.19 [$18-20] **Wedding Ring Cloche** with feather stick-up, grosgrain band on wool felt. $2.49 [$18-20] **Cute as a Button Helmet** in 100 percent wool felt cloth. $1.79 [$18-20] **Beautiful Bonnet,** always new, adds double quills, grosgrain band. Wool felt. $2.69 [$18-20] **Storm Cap** in sturdy cotton gabardine has mouton lamb earflaps. Warm quilted rayon lining. $2.59 [$18-20]

Peak Cap of knitted wool for cozy warmth, bright color. Gay stripe accent $2.69 [$20-22] **Adorable Roller** in cotton velveteen and grosgrain band, bow. Elastic at back. $2.98 [$20-24] **Genuine Basque Beret** in wool. $1.14 [$15-18] **Luxury-rich Laskin** mouton lamb. A warm fringe on knitted wool helmet. Head clip for snug fit. $2.89 [$18-20] **Roller Derby** spiced with grosgrain band, multicolored rayon taffeta bow. Wool felt. $2.64 [$18-20]

Becoming Bonnet. Scoop-brim adds extra width, lustrous ornaments harmonize on wool felt. Rayon cage veil ties in back. $3.98 [$18-22] **Pretty Petal Brim** with rich rayon velvet is a fashionable accent on fine wool felt off-face hat. Twin swirling feathers, self-band. Flattering rayon veil. $3.98 [$30-35]

One of Our Finest wool felt hats is tiered for a new variation. Rhinestone ornament, scallop-edge self-wings. Enchanting rayon veil. $3.98 [$24-28] **Important Off-face** silhouette fashioned in wool felt with rayon velvet facing and fluffy side feather. Rayon veil. $3.79 [$28-32] **Tiered Platform Hat** in wool felt and chenille. Rayon cage veil with wool felt tie streamers. $3.98 [$24-28] **Luxurious Feathers** atop the popular off-face hat. Rayon velvet band, wool felt, rayon veil. $3.98 [$30-35] **Brow-praising Bonnet** with feathers, gold-color touched stem. Wool felt. Rayon veil. $3.98 [$30-35]

Ostrich Feather Plume on wool felt homburg. Crisp rayon veil. Colors: $3.69 [$35-40] **Off-face Hat.** Gently curving feathers accent fine wool felt. $3.98 [$24-28] **Side-dipped Hat** in always-right wool felt, fluffy feather bouquet, self-bow, and rayon veil. $2.98 [$30-35] **Roll-edge Sailor** in fine wool felt. Rhinestone with rayon velvet touches, fan swirl, and band. Rayon veil. $3.98 [$24-28] **Pearl Highlighted Beret.** Wool felt. Self-trim winging through pearl-like trim. $3.98 [$24-28]

Poke Bonnet in rayon velvet with two self-covered buttons. Chenille dotted rayon veil. $3.49 [$20-22] **Petal Off-face Brim.** Feather on wool felt hat. Self band, bow. $3.98 [$28-32] **Prized Pinwheel Feathers** on rayon velvet close-clinging helmet. $2.69 [$28-32] **Wedding Ring** in wool felt. Double quill, self bow. $2.69 [$28-32] **Angora Cloche,** 70/30 wool/cotton stripping with gold-color band, side feather. $3.39 [$20-24]

Cute-as-a-button Helmet in rayon velvet. Self-covered buttons. $2.98 [$20-24] **Rayon Velvet** covered buttons on felt cloth (70/30 wool/cotton). Rayon veil. $2.98 [$20-24] **Perky Pill Box** in wool felt with twin rhinestones. Rayon cage veil, self-streamers. $3.29 [$18-20] **Gleaming Bumper** in wool felt with glittery band. Rayon veil.. $2.98 [$20-22] **Rayon Velvet** pill boxed with medallion pin. $2.98 [$18-20]

Derby Winner in wool felt, rayon satin band, bow. $2.69 [$20-22] **Basque Beret** in velour finished wool, imported from France. $1.49 [$15-18] **Dutch Cap.** In cotton velveteen. $2.19 [$20-24] **Jockey Cap.** Grosgrain trim. Cotton velveteen. $2.19 [$20-22] **Pearl-like Buttons** on helmet. Cotton velveteen. $2.19 [$20-24]

If head measures	Order size
21¼ in.	6¾
21⅝ in.	6⅞
22 in.	7
22⅜ in.	7⅛
22¾ in.	7¼
23⅛ in.	7⅜
23⅝ in.	7½

Note: The fine fur felt hats for men tend to bring higher prices than the wool felts. When men's hats are available at estate auctions, prices tend to be low, only a few dollars per hat. By the time they make it up in the ladder to the clothing shows, prices escalate tremendously.

Southwest Flight style. Brim is 2.75 inches wide. Fine quality fur felt; crown can be shaped as you like. Form Ease sweatband. $6.95 [$50-55]

Pilgrim Fur Felts. Three styles with full rayon lining. Rayon band, trim. Pliofilm tip in crown to resist perspiration stain. Leather sweatband. 2.5-inch brims. Fur felt: Ambassador style. $4.98 [$35-45] **Southwest Flight Style.** $4.98 [$35-45] **Pathfinder** style. Plain edge brim. $4.98 [$35-45]

Ambassador Form Ease. Dressy style, with ribbon-bound edge on the 2.75-inch brim. Crown is pre-blocked at the factory. Band is in color contrasting with color of hat. Fine fur felt. Patented Form Ease sweatband. $6.95 [$50-55]

Warm 'Coonskin Pioneer-style Cap. Made of raccoon pelts. Reflector "eyes" gleam in the dark. Water-repellent capeskin crown. Turndown in-band of 100 percent wool. Lining of quilted rayon. $2.94 [$75-95]

Note: The market for young boys' hats from the early 1950s is not much stronger than it is for girls' hats. Some exceptions would be the "helmets with goggles" sets, and the racoon fur pioneer-style caps which are considered collectible and interesting for display. When the movie Davey Crockett was released in the early 1950s, both boys and girls longed for the pioneer style cap that Davey wore, and it became a very popular style for some time.

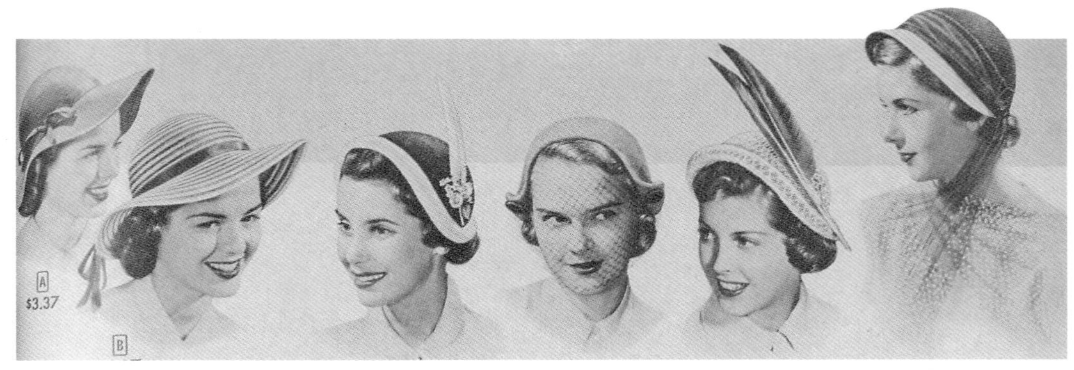

Flowery Brimmed Casual. Shape brim to suit your moods.Wool felt. $3.37 [$28-35] **Wedding Ring** in straw-like cloth. Flower-feather fancy. $2.59 [$20-22] **Dutch Treat** with rayon velvet accents on covered buttons and tie streamers on rayon cage veil. Straw-like cloth. $2.29 [$20-22] **Twin Quills** on your favorite wedding ring. Starched white cotton boucle. $2.39 [$22-24] **Becoming Bonnet** with contrasting brim, lavish chenille dotted rayon veil. In straw-like cloth. $3.49 [$28-35]

Face Framing Feathers, a luxurious fashion extra on halo half hat. Straw-like cloth. Rayon veil. $2.29 [$20-22] **Shell Bonnet** with floral wreath. Straw-like cloth. Rayon veil. $2.98 [$24-28] **Helmet Beauty.** Hand-crocheted straw braid. Fluffy pinwheel feather with harmonizing flower center. $3.49 [$28-35] **Popular New Shell.** An exciting fashion favorite traced with rayon braid. Rayon tie-back veil. Straw-like cloth. $2.98 [$18-20] **Spring-fresh Blossoms** on straw-cloth bonnet. Rayon veil. $2.59 [$28-30]

Honey of a Helmet with feather-fancy and grosgrain bow. In straw-like cloth. $2.29 [$24-32] **Flower Show.** A garland of harmonizing flowers on expertly draped hat. In straw-like cloth. Rayon veil. $2.59 [$18-20] **Flower Festival.** Colorful flowers on halo half-hat. In glimmering, cellophane braid. $2.59 [$35-45] **Tricorne Triumph** stars a sweetheart cuff, top button, lavish rayon veil. In straw-like cloth. $2.59 [$24-30] **Glamour Calot.** Straw braid stitched in leaf design. Chenille dotted veil. $2.39 [$28-35]

Dashing Cavalier Hat. Straw-like cloth that features a lavish feather, rayon veil. $3.98 [$35-38] **Angel-faced Bonnet.** Harmonizing flowers, rich rayon taffeta tubing, secured rayon veil, all on straw braid. $3.98 [$40-45] **Sweetheart Hat.** Two-toned double tier in fine straw braid. Nylon veil with chenille dotted border. $3.98 [$40-45]

Feathered Flatterer. Curled feathers hand-pasted on a helmet that dips to one side. $3.98 [$28-32] **Style-worthy Shell** with velvet accents, harmonizing flower garland, twin bows, band on fine braid. Rayon veil. $3.98 [$24-28]

Cloche Allure. Rhinestones on nylon veiling. In straw-like cloth. $4.98 [$35-40] **Important Little Hat.** Harmonizing floral clusters, rayon velvet bows, woven straw braid. Rayon veil. $4.69 [$18-20]

Crowning Glory floral wreath on straw braid hat. Rayon velvet bows, rayon veil. $3.98 [$18-20] **Versatile Picture Hat** in straw braid. Grosgrain band, bow. $3.98 [$28-35]

On-the-level Shell Success. Fancy woven straw braid with blending flowers, rayon velvet trim, rayon veil. $4.69 [$24-28] **Candy Straw Delight.** Rayon taffeta tubing outlines the unusual crown and beautiful chenille-dotted rayon veil. $3.69 [$18-20] **Picture-pretty Hat.** Brim forms a flattering fashion circle. Rayon velvet highlights. Woven straw imported from Italy. Rayon veil. $3.59 [$28-35]

Off-face Silhouette. Straw braid. Fluffy pinwheel feather, rayon velvet band. Rayon veil. $3.98 [$28-32] **Flower Show-off.** Embossed rayon leaves, misty rayon veiling. $4.98 [$24-28]

Square-brim Bonnet. Harmonizing floral wreath, rayon velvet bows, rayon veil on straw braid. $$3.98 [$18-20] **Charm of a Helmet** blending floral clusters, rayon velvet tubing, rayon veil. In candy straw. $3.98 [$24-28]

Glamorous Sweep of a deep off-face brim, blending flowers and rayon velvet band on fine straw braid. $.98 [$28-32] **Poke Shell.** Floral wreath on rayon velvet band and twin perky bows. In fine straw braid. Rayon veil. $3.98 [$24-28]

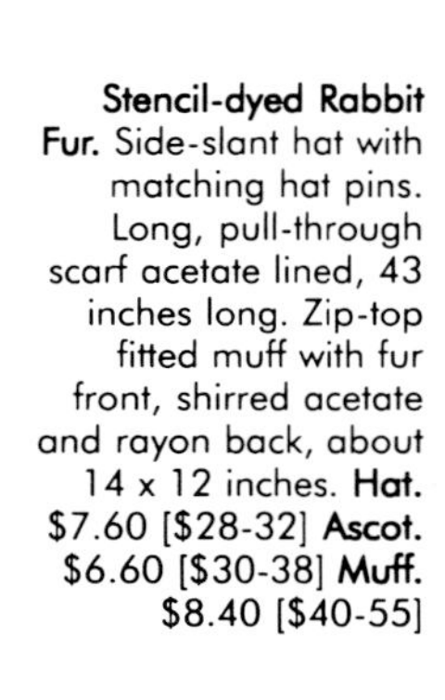

Stencil-dyed Rabbit Fur. Side-slant hat with matching hat pins. Long, pull-through scarf acetate lined, 43 inches long. Zip-top fitted muff with fur front, shirred acetate and rayon back, about 14 x 12 inches. **Hat.** $7.60 [$28-32] **Ascot.** $6.60 [$30-38] **Muff.** $8.40 [$40-55]

Note: Fur has a very erratic reception in the vintage clothing market. Many dealers will not sell fur at all, others, realizing the high cost of repair and storage, are reluctant to buy fur unless it is in perfect condition and looks as if it will sell quickly. Basically, the anti-fur trend hit the vintage market before the larger market. As a dealer, buy carefully. As a collector, buy very carefully because many of the vintage coats available have not been given proper fur storage and may not hold up.

Rayon Velvet with silver-color accented self-quill. Rayon veil. $3.98 [$40-45] **Ostrich Feathers** on rayon velvet profile cuff. Wool felt crown. Rayon veil. $4.98 [$40-45]

Side-dipped Hat in wool felt. Feather spray, rayon veil. $2.98 [$28-32]

Off-face Loveliness. Mirror-stemmed feathers on double-tiered brim. Rayon velvet with compliment-catching rayon veiling. $3.59 [$28-32] **Dress-up Homburg** in rich rayon velvet. Feathers and rhinestones. Grosgrain band. Rayon veil. $3.98 [$30-35] **Graceful Face-praise.** Natural ostrich feather aloft off-face wool felt hat. Self-band. Rayon veil. $3.98 [$30-35]

Sailor Flattery in wool felt. Rayon taffeta tubing peeks through double edge brim. Grosgrain band, bow. Draped rayon veil. $3.99 [$20-22] **For the Gracious Lady.** Lovely ostrich feather on wool felt hat. Rayon veil. $2.98 [$28-32] **Halo Bumper.** Grosgrain cockade with a jeweled center. Wool felt. Rayon veil. $3.98 [$18-20] **So Much Beauty, So Much Style.** Rayon velvet band, fluttery feather, detailed crown. Rayon veil. $3.29 [$35-45]

***A:* Ecuadorian Panama.** Front pinch shape. Snap brim. Hook-on sash band of rayon crepe. Roan leather sweatband has oiled silk underlining. $6.50 [$40-45] ***B:* Ecuadorian Panama.** Snap brim. Roan leather sweatband. $4.50 [$45-50] ***C:* Toyo Bangkok,** low-priced favorite. Side ventilation. Snap brim. DuPont imitation leather sweatband. $1.98 [$25-30] ***D:* Imported Fiber.** Side walls ventilated. Perforated roan leather inner band. Featherweight rayon mesh lining. Snap brim. $2.95 [$40-45]

Note: The market for men's felt hats was stronger several years ago than it is now. The trend to wear men's felt hats may be inspired by the release of a movie, or a celebrity presence wearing such a hat (as Harrison Ford in Indiana Jones and the Temple of Doom).

Lovely Straw Braid Hat with harmonizing flowers, grosgrain trim. $2.59 [$30-35] **Dressy Homburg.** A harmonizing floral cluster and grosgrain trim on straw braid. $2.98 [$40-45] **Two-tone** grosgrain ribbon on straw braid. $2.59 [$45-50]

***Above left to right:* Shell Bonnet.** Floral wreath spices straw-like cloth. Rayon veil. $2.98 [$18-20] **Daisy Shell** around straw-like cloth, tie-back rayon veil. $2.98 [$18-20] **Cuffed Profiler** in straw-like cloth with a spray of flowers. $2.29 [$24-28] **Lavish Rayon Veiling,** plus rayon velvet accents and harmonizing floral clusters on straw-like cloth bonnet. $2.29 [$24-28]

***Left top to bottom:* Tucked Wedding Ring** adds a big rose. Rayon veil. In straw-like cloth. $2.59 [$28-32] **Rayon Soutache Braiding,** pleated grosgrain cockade on straw-like cloth helmet. Rayon veil. $2.29 [$18-20] **Roof-top Bonnet** with yellow-centered daisies. In straw-like cloth. Rayon veil. $2.98 [$20-22] **Glamour Halo** of rayon veiling, flecked with crisp dots, circles straw-like cloth bonnet. $2.29 [$18-20]

Two-tone Grosgrain in straw braid with imitation pearl ornaments, grosgrain trim. $2.39 [$30-35]

Picture Hat. Superbly simple. In fine straw braid. Grosgrain band, bow. $3.98 [$40-45] **Big-bow Picture Hat** in lacy straw braid with rayon taffeta band, bow. $3.49 [$40-45]

Umbrella Bonnet with a rippling, wavy brim and floral wreath. In straw cloth. Tie-back rayon veil. $3.29 [$20-22] **Flowering Shell.** Lilacs with rhinestones on a straw-like cloth shell. Tie-back rayon veil. $2.98 [$20-22] **Embroidered Treat.** Lavished with embroidered French knots. Toyo straw. Tie-back rayon veil. $3.49 [$30-35] **Draped and Shapely** brim, with a garland of flowers, on a bonnet. Rayon velvet bows, rayon veil on straw braid. $3.98 [$20-22]

Profile Drama. Artfully manipulated straw braid hat with a curved pleated grosgrain stick-up, imitation pearls, rayon veil. $3.98 [$20-22] **Sailor Loves Flowers.** A wreath of flower, rayon velvet band, rayon veil. $3.98 [$20-22] **Make-you-pretty Profile.** Straw braid beret with a bouquet of country flowers and rayon velvet bow. Rayon veil. $3.98 [$20-22] **Two-Tone Sailor.** Contrasting edging and two-tone triangles center grosgrain bow on sisal straw. Rayon veil. $3.98 [$28-32]

Flower and Veiled Allure. Rhinestones on nylon veiling that covers flowers. Straw-like cloth. $5.98 [$30-35] **Lilies-of-the-valley** border a flattering shell with rayon velvet bows, rayon veil. Straw-like cloth. $3.98 [$22-24]

Profile Jewel. Imitation pearls and rhinestones on straw braid. Wired rayon velvet loops. Rayon veil. $3.98 [$30-35] **Profile Beret.** On one side a floral rosette, and a rayon velvet band follows the curve. In straw braid. Rayon veil. $3.98 [$28-32] **Prize for Flowers.** Lilacs border profile helmet. Rayon velvet trim, rayon veil. In toyo straw. $3.98 [$20-22] **Flower Show Bonnet.** Embossed rayon leaves, a border of shy blossoms, rayon veiling. $3.98 [$20-22]

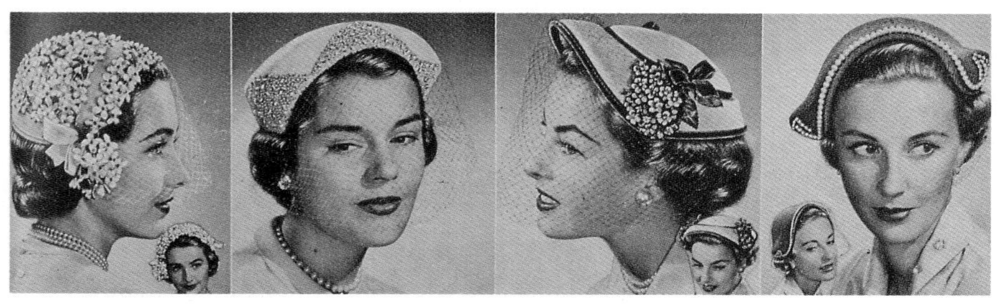

Glorious Helmet. You're wearing a make-believe garden, alive with the dewy glitter of rhinestones and under-scored by vibrant rayon velvet touches. $3.98 [$28-32] **Pill Box Treasure.** Beads and other sparklers shine night and day on toyo straw. Rayon veil. $3.98 [$18-22] **Bonnet Shell.** Flower clusters, rayon velvet extras, on toyo straw. Tie-back rayon veil. $3.98 [$18-22] **Profile Pearl.** Imitation pearls and rhinestones on straw-like cloth. Rayon veil. $3.98 [$18-22]

A: Comfortable Classic. Grosgrain trim, rhinestones on wool felt. $2.98 [$22-24] **B: Great Profiler.** Natural ostrich feather sits high on wool felt. Back rayon veil. $2.98 [$24-28] **C: For the Gracious Lady.** Ostrich feather. Wool felt. Rayon veil. $2.98 [$24-28] **D: Winging Profile.** Skyward-bound feather, glitter-splashed on draped wool felt beret. $2.98 [$20-22] **E: Perfect Profiler** with pinwheel feather and rolled cuff on wool felt. $3.49 [$24-28] **F: Feathered Shell.** Rhinestone dotted. Tie-back rayon veil. $2.98 [$24-28] **G: Metallic** embroidery on double-tiered wool felt shell. Tie-back rayon veil. $3.49 [$24-28] **H: Sorcery in Sequins** helmet. Tie-back rayon veil. $3.49 [$24-28] Bonnet Supreme. A curling feather anchored by sham pearls. $3.49 [$24-28] **J: Bonnett Supreme.** Rayon velvet with curling feather anchored by sham pearls. $3.49 [$24-28]. **K: Golden Touched Fuzzy Cloche.** Angora rabbit hair on wool felt, banded with gold-color sequins. $2.98 [$20-22] **L: Icing on the Shell.** White beading on rayon velvet shell. $3.49 [$18-20] **M: Heavenly Helmet.** Bead-sprinkled rayon velvet tubing with sheer rayon net. $3.49 [$20-22] **N: Profiler,** with rhinestone-stemmed feather, braiding on wool felt. Rayon veil. $2.98 [$24-28] **P: Fanciful Feather Spray** aloft rayon velvet hat with pleated off-face cuff. $2.98 [$24-28]

Halo Cover-up. Suede rayon. $1.59 [$32-35] **Crowning Cover-up.** Glitter sprinkled on rayon velvety chenille and metallic cuff. $1.98 [$28-32]

Fashion Charmer. Wool felt with sham pearls, tie-back rayon veil. $2.59 [$18-20] **Fashion Winner** with rayon velvet cuff, fluffy feather on wool felt. $2.39 [$20-22] **Twin-quilled Bonnet.** Wool felt. $2.49 [$24-28]

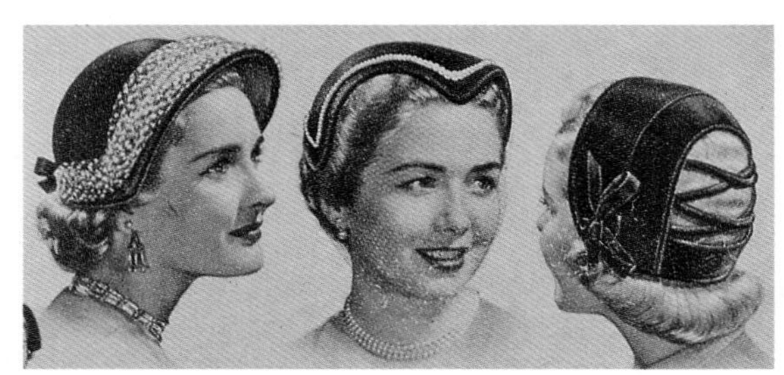

Veiled Halo. Chenille dotted rayon veiling on rayon velvet bonnet. $2.49 [$18-20] Sweetheart Shell. Sham pearls on a rayon velvet shell with tie-back rayon veil. $2.49 [$18-20] **Laced-back Calot.** Rayon velvet. $1.98 [$24-28]

Calot Charmer. Glitter on perky bows and flowers, all around a wool felt calot. $1.79 [$28-32] **Highlights on a Pill Box.** Imitation pearls on rayon velvet cuff frame wool felt crown. Rayon veil. $2.59 [$18-20] **Feminine Success.** Wool felt spiked with a curling feather. $2.49 [$18-20]

Ruffled Shell Set in rayon velvet. Rayon veil. Petal drawstring bag (7 x 6 inches) is rayon lined. $5.98 set [$40-45] **Shell Treasure** in rayon velvet, sham pearls, rhinestones, and rayon veiling. $3.98 [$18-20] **Shimmer Calot.** White beading on rayon velvet. Rayon veil. $3.98 [$20-22] **Row on Row** of rayon braiding outlines a rayon velvet bonnet. Rayon veil. $3.98 [$20-22] **Fluffy Feather** clusters and rhinestones on rayon velvet helmet. $2.59 [$24-28]

Array of Sham Pearls on a grosgrain band. Rayon velvet bonnet. $3.98 [$20-22] **Pill Box** in rayon velvet. Tie-back rayon veil. $3.49 [$20-22] **Shell** sparkles with sham pearls and is rayon velvet. Rayon veil. $2.49 [$20-22] **Glitter Row** of delicate sham pearls and other sparklers on rayon velvet helmet. $3.49 [$20-22] **Feminine Affair.** Off-face rayon velvet hat with frilly feathery gusher. $3.59 [$22-24]

Mirror-stemmed Feather and rayon veil spice rayon velvet hat. $3.79 [$24-28] **Frosted Pill Box.** White beads ice rayon velvet hat. $4.49 [$22-24] **A Feather Fancy** on a rayon velvet off-face flatterer. $3.49 [$32-34] **Crowning Pill Box.** Rayon velvet hat topped with rhinestones. Tie-back rayon veil. $3.98 [$18-20] **Rayon Velvet Hat** flaunts a feather fancy. $2.79 [$24-28]

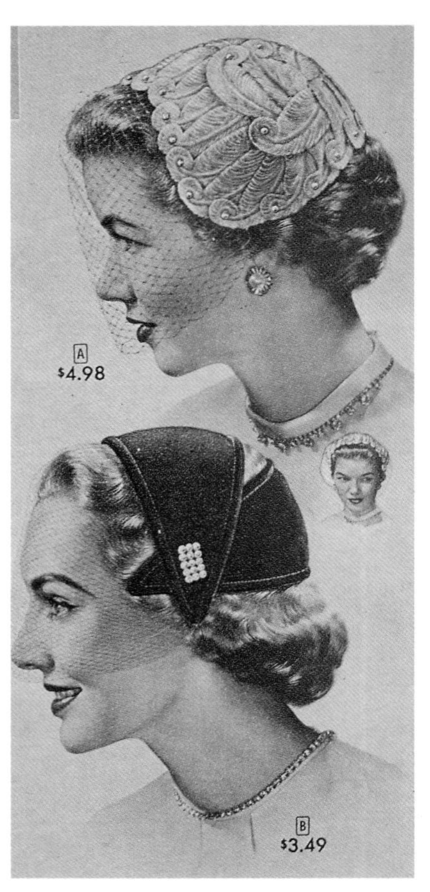

Velvet Feather Delicacy. Rayon velvet embossed leaves are magically treated to give the illusion of ostrich feathers, then sprinkled with rhinestones. Rayon veil. $4.98 [$22-24] **Romantic Cap.** Rayon velvet with rhinestones and tie-back rayon veil. $3.49 [$22-24]

Winter Flowers with rhinestones on rayon velvet bonnet. $4.49 [$18-20] Curving Coque, rayon velvet, rayon veil. $4.98 [$24-28] **Fall Fireworks.** Glittering fashion arrow lights up rayon velvet profiler. Rayon veil. $3.98 [$24-28]

Shutter Cap. Dazzling nail-heads and a sprinkling of rhinestones on rayon velvet. $2.69 [$24-28] **Crowning Shell.** Sparklers on rayon velvet with rayon veil. $3.98 [$20-22] **Lady-finger Calot.** Sham pearls and beading glow on rayon velvet. Tie-back rayon veil. $3.98 [$24-28]

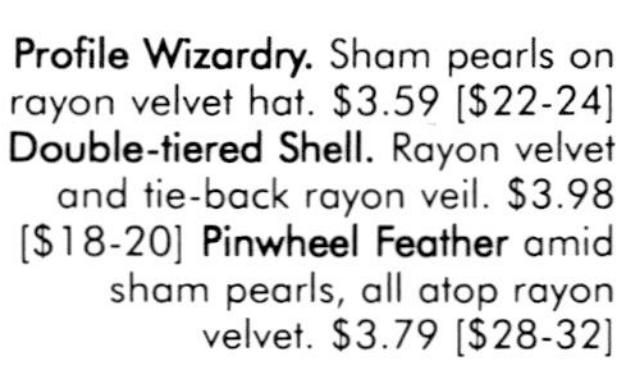

Profile Wizardry. Sham pearls on rayon velvet hat. $3.59 [$22-24] **Double-tiered Shell.** Rayon velvet and tie-back rayon veil. $3.98 [$18-20] **Pinwheel Feather** amid sham pearls, all atop rayon velvet. $3.79 [$28-32]

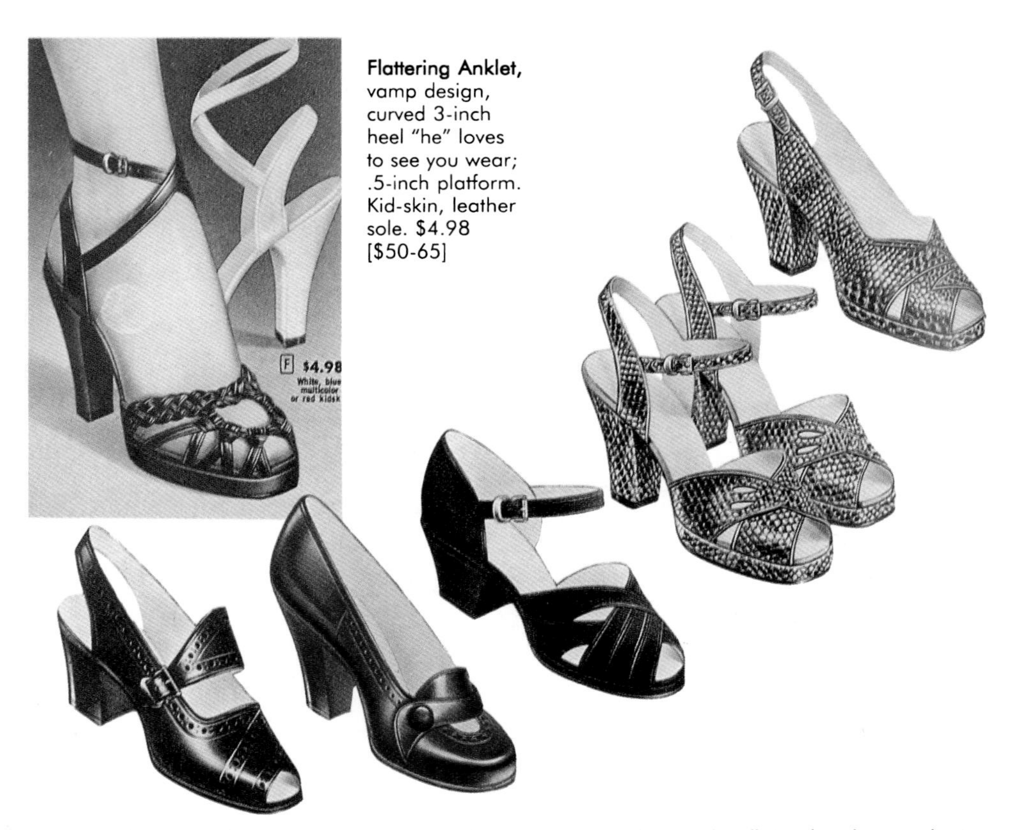

Flattering Anklet, vamp design, curved 3-inch heel "he" loves to see you wear; .5-inch platform. Kid-skin, leather sole. $4.98 [$50-65]

Left to right: **Trim Style** with broad adjustable sabot strap, square throat. Like all Kerrybrookes, quality is guaranteed. Leather sole, 1.75-inch heel. $4.98 [$35-40] **Dainty Pump.** Covered heel to toe. Perforations dot vamp, collar. Leather sole, 2.25-inch heel. $4.98 [$40-45] **Softly Draped Vamp.** Open toes and sides. Leather sole, .25-inch platform, 1.5-inch curved Louis heel. $2.98 [$45-50] **Foot-flattering Sandal.** Diagonal vamp bands, instep strap adjusts. Leather sole, .5-inch platform, 2.5-inch heel. $3.98 [$50-65] **Envelope Vamp,** V-throat, .5-inch platform. Adjustable strap, leather sole, 2.5-inch heel. $3.98 [$50-65]

Note: In the early 1950s some of the most collectible of the vintage shoes were in style—notably the low-platform sling-back pump, the low-platform open-toe pump, and the low-platform crossed-ankle strap sandal. Sears offered some of the most incredible styles of the period in quality materials. Designers of accessories are fascinated with vintage footwear of all periods. Women collectors love to buy and wear these beautiful styles, which are often in very good condition and offered at affordable prices compared with the current designer equivalents, which are priced upwards of $175.

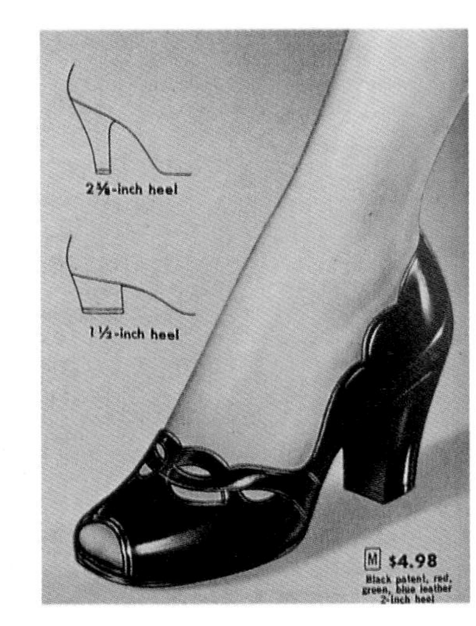

Patent or Leather Pump. Scalloped collar. Leather sole. $4.98 [$45-50]

Chapter 5

Shoes and Socks

Top right to bottom left: **Bracelet Sandal.** .5-inch platform, 3-inch heel, leather lined and soled. $7.95 [$65-75] **Sling Pump, Open Vamp,** .25-inch platform, 2.75-inch heel, Kerrybrooke. $6.98 [$65-70] **Siren Bracelet Sandal,** V-throat, lacy vamp openings, .5-inch platform, 3-inch heels, leather sole. $6.98 [$65-75] **Curved Straps, Openly Cool,** .25-inch platform, 2.5-inch heel. $7.95 [$65-75] **Sling Pump.** Curved throat, .5-inch platform; slender 3-inch heel. $7.95 [$65-75] **Suede, Patent, or Calf.** Tiny vamp openings, .25-inch platform, 2.25-inch heel, leather sole. $5.98 [$65-75] **Sweet Bracelet Sandal,** .5-inch platform, 2.5-inch heel. $6.98 [$65-75]

Foot-slimming, Size-Deceiving Sandals. Open toe and back, built up sides for support. California-style cushion comfort platform and wedge heel. $4.98 [$45-50]

Sweet and low

Top: **Western Touch.** Hand-tooled, hand-laced leather. Soft platform, wedge heel, composition sole. $3.98 [$85-90] **Lovely Pump.** Knotted crisscross straps over the vamp, baby doll toe, low wedge heel. Adjustable sling strap. $5.98 [$65-75]
***Left:* Ladylike Pump.** Oval ornament on the vamp. Leather sole, flat heel. $5.98 [$55-60] **Easy to Polish.** Contrasting piping on the vamp. Leather sole, heel. $5.98 [$45-50] **Peaked Back Casual.** Searosole, wedge heel. $3.98 [$65-70]

Top left to botton right: **Wide-width Shoes: Dressy Sandal.** V-throat, cool vamp openings; adjustable instep strap. Platform, wedge heel. $3.98 [$55-60] **T-Strap Sandals.** Platform comfort, wedge heel. $4.98 [$45-50] **Slip-on Casual.** Rounded vamp, cushion platform, wedge heel, Searosole. $3.98 [$65-70] **Twin-strap Sandal.** Oval vamp openings, open toe, open back. California-style platform and wedge heel. Searosole. $2.98 [$65-70] **Open U-throated Tie,** 5-eyelets, platform, wedge heel. Searosole. $3.98 [$65-75] **Tie That Feels Good.** Platform, wedge heel, Searosole. $3.98 [$65-70]

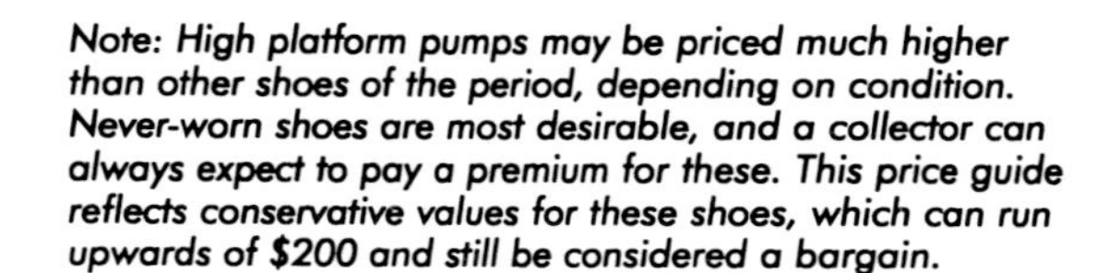

Note: High platform pumps may be priced much higher than other shoes of the period, depending on condition. Never-worn shoes are most desirable, and a collector can always expect to pay a premium for these. This price guide reflects conservative values for these shoes, which can run upwards of $200 and still be considered a bargain.

Lerft to right: **Leather Slip-on.** Platform, wedge heel. $3.98. [$65-70] Extra Support. Cushion-soft platform, wedge heel. $3.98 [$65-70] **Covered Back, Open Toe,** platform, wedge heel. $3.98 [$65-70]

F: **Airy Sandal.** T-strap and adjustable in-step strap. Platform and wedge heel. $3.98 [$45-50] ***G:*** **Slenderizing V-line Vamp.** Swirl strap, platform, wedge heel, Searosole. $3.98 [$40-45] ***H:*** **Round-the-clock Sandal.** V-throat, adjustable instep strap, platform, wedge heal. $3.98 [$40-45] ***J:*** **Pretty Braided Vamp.** Cushion-platform, wedge heel, buckled strap, Searosole. $3.98 [$40-45] ***K:*** **Clever Swirl Strap.** Envelope vamp, cushion-platform, wedge heel, Searosole. $3.98 [$45-50] ***L:*** **Leather Sling Pump.** Circle vamp ornament, platform and wedge heel. $3.98 [$55-60] ***M:*** **V-throat Vamp.** Adjustable strap, platform, wedge heal. $3.98 [$45-50] ***N:*** **Bracelet Sandal.** Snakeskin, or smooth leather, platform, high wedge heel. $3.98 [$55-60] ***P:*** **"Braided" T-strap Sandal.** California-platform, wedge heel. $3.98 [$45-50] ***R:*** **Leather Strap Sandal.** Platform, wedge heel, Searosole. $3.98 [$45-50]

Left to right: **Wishbone Sandal.** Perforated vamp and strap, platform, wedge heel. $2.98 [$35-40] **Pool-cool Leather Sandal.** V-throated vamp, platform, wedge heel. $2.98 [$35-40] **U-throated Tie.** Laces high at instep, teardrop eyelets. $2.98 [$65-75] **Flexible Leather Sandal.** Vamp draped to look like a bow. $2.98 [$45-50] ***Top right:* Foot-coddling Leather Sandal.** Cushion-platform, wedge heel. $2.98 [$45-50]

Left to right: **Cotton Duck Sandal.** Braid vamp, elasticized ankle strap, crepe-like rubber sole and wedge heel, rope design tread. $1.98 [$30-35] **Cotton Duck Oxford.** Imitation suede trim, crepe design rubber sole, 1.75-inch heel. $1.98 [$65-70] **Three-eyelet Oxford.** Ribbed cotton, foam rubber platform, wedge heel, and sole. $1.98 [$30-35]

Note: Early television programming glorified the cowboy and the Old West. Many items of clothing for children reflected this motif. The cowboy items go beyond vintage wearable and into the collectible arena. Very likely such boots as the ones pictured would end up as "shelf shoes," reminiscent of childhood. Otherwise rubber boots have little value, except to costumers, or if they have a particularly interesting design for a fashion designer trying to create a "new" look!

Fellows! Girls! Cowboy hero Roy Rogers and his horse Trigger in multicolor design on an all rubber over-shoe boot. Have Mom order your pair today. $3.29 [$85-90]

Not Just a Toy but a sturdy, waterproof rubber over shoe boot that you won't have to coax the kiddies to wear. $2.89 [$85-90]

Women's Kerrybrooke Storm Rubbers made extra wide. $1.59 [$15-20] **Toe Rubbers.** Easy on-off. $.95 [$10-12] **Storm Rubber.** Lined. $1.49 [$15-20] **For Girls, Children.** Flexible rubber. Extra layers at wear points. $1.89 [$15-20] **Kerrybrookes** for women, misses. $1.98 [$15-20]

Gypsy Tie for all day comfort. Kid with patent trim. Leather sole, 1.75-inch heel. $7.95 [$65-70]

Slip-on with decorative vamp cut-outs, cool open back, adjustable sling strap. 2-inch heel. $7.95 [$70-75] **Five-eyelet Nurses' Style** kidskin oxford, leather sole, 1.5-inch heel. $7.95 [$55-60]

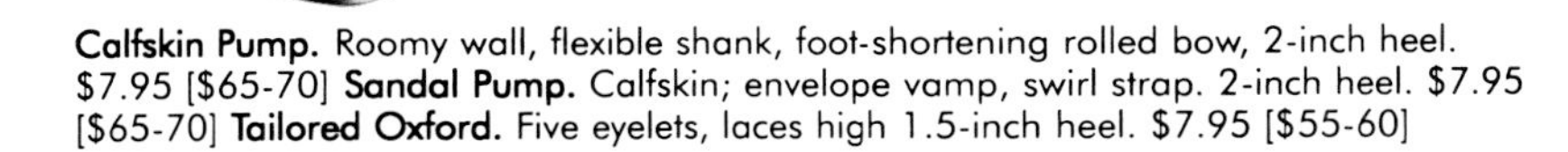

Calfskin Pump. Roomy wall, flexible shank, foot-shortening rolled bow, 2-inch heel. $7.95 [$65-70] **Sandal Pump.** Calfskin; envelope vamp, swirl strap. 2-inch heel. $7.95 [$65-70] **Tailored Oxford.** Five eyelets, laces high 1.5-inch heel. $7.95 [$55-60]

Twin-strap Sandal. Pliant platform, wedge heel. $5.49 [$55-60] **Foam Rubber Insole,** heel pad, arch cushion. Platform, wedge heel. $5.49 [$50-55] **Tie** for extra comfort. Foam rubber cushion. $5.69 [$35-40]

Clockwise from top: **Leather Saddle Strap Slip-on.** Rounded toe, back. Foam rubber arch cushion, platform, low wedge heel. $5.49 [$40-45] **Customer Favorite.** Casual blucher tie. $4.98 [$40-45] **Six-eyelet Tie.** Platform, wedge heel. $3.98 [$65-70] **Four-eyelet Tie.** Tear drop vamp openings. $4.98 [$65-70]

Arch-supporting Foam Rubber Cushion. Plump platform, wedge heel, $5.49 [$65-70]

Foot-slimming Swirl Strap. V-throat, envelope vamp. $2.98 [$60-65] **Narrow Triple Straps,** widely spaced. $3.49 [$60-65]

Go-with-everything Sandal. Twin-pointed vamp, V-throat line. $3.49 [$45-50] **Feminine** pump. Low cut vamp, foot-shortening bow, sling strap. $3.98 [$65-70] **Smart Bow Pump.** Squared toe, snug heel strap. $2.98 [$65-70] **Wishbone Sandal.** Easy walking platform, 2.25-inch wedge heel. $3.49 [$65-70]

Clockwise from top: **Bracelet Sandal,** leaf-shaped openings. $3.49 [$55-60] **Slip-on Softie.** Pinked around the mock vamp and peaked tongue, two buckled instep straps. $2.98 [$40-45] **Tailored Sing Pump.** Circular vamp ornament. $3.49 [$60-65] **Fun-time Sandals.** Braided vamp bands, adjustable strap. $2.98 [$50-55]

Note: Tie-up oxford styles have become increasingly interesting to collectors in recent years. Women who wear 1940s and 1950s wool gabardine suits like the look. During the 1950s, the tie-up styles were most popular with older women. My mother preferred wearing the open toe high-heeled pump. My grandmother had the oxford shoes in several dark colors and white, often ordering from Sears and considering these shoes "as fine as any shoe in Boston."

Dressy Slip-on. Elastic side gore, 2-inch heel, rubber lift. Black. $5.98 [$65-70] **Feminine Tie** in glowing calfskin, 2-inch heel. $5.98 [$70-75] **Crushed Kidskin** with patent, 2-inch heel. $5.98 [$70-75] **Sturdy, Flexible Goatskin Tie.** 1.5-inch heel, rubber lift. $5.98 [$55-60]

Gypsy Oxford. Leather sole, 1.75-inch heel, rubber lift. $7.65 [$65-70] **Twin-strap for Smart Styling.** High at sides and back; low curved throat, dainty vamp openings, 2-inch heel. $7.85 [$65-70]

Above: Moc-toe Gone Western for your ramblin' pleasure. Embossed "steer-horn" on vamp, flashy buckle and strap. $6.65 [$85-110] **Custom-toe Oxford.** $6.65 [$20-25] **Burly Brown Leather Upper,** roomy wall toe, massive seam around the front. $6.65 [$30-35] ***Right:* Smart Plain-toe.** Leather upper. Goodyear welt sewn. $6.65 [$20-25] **Rugged Lines** plus medallion punch work on the roomy wall toe. $6.65 [$20-25]

Note: Men's clothing and accessories have remained traditional when compared with women's fashions and accessories. Shoes that are particularly unusual, such as the embossed leather "buckle and strap" type, are collectible. The more traditional styles have little or no market value since similar styles are being made today. Vintage alligator shoes for men are always very desirable.

Roomy moc-toe, raised seam. $5.49 [$20-25] **Steerhorn Design** embossed on vamp, flashy buckle and strap, roomy moc toe. $5.95 [$85-110]

Note: There is a definite market for vintage bedroom slippers, especially the beautifully made satin slippers from the early 1950s. Many styles were made of quilted rayon satin. Collectors enjoy wearing the slippers with robes of the period, lounging pajamas, and sometimes even with evening gowns.

***G:* Lustrous Rayon Satin Honey,** topped with luxurious white fur cuffs. $1.98 [$35-40] ***H:* Quilted Rayon Satin.** Ridged moc-style vamp, tiny bow. $1.98 [$28-32] ***K:* Cotton Velveteen,** all-over chenille effect rayon satin platform. $2.19 [$30-35] ***L:* Paris-inspired** scuffs, rayon crepe, embroidered in rose design, 1.5-inch wedge heel, leather sole. $3.79 [$28-30] ***N:* Luxurious Scuff** in sculptured rayon satin, white fur. $2.19 [$35-45] ***T:* Smooth Rayon Satin Scuff.** Shirred vamp, stand-up frill at the throat. $2.35 [$30-35] ***V:* Scalloped-throat Scuff.** Rayon satin, diamond vamp design. $2.25 [$30-35] ***W:* Rayon Satin Slippers.** Open toe, embroidered throat. $2.19 [$20-25]

Note: Saddle shoes were popular school shoes during the early 1950s. The colors available were white with black saddles and white with brown saddles. Although the vintage market for these shoes is not strong, they are still being manufactured with minor differences from the originals, the older shoes tend to have a slightly different look than contemporary copies.

Scientifically Designed and progressively improved. Strong, flexible Goodyear welt construction. Drill cloth vamp lining, leather heel lining and insole for inner smoothness, greater strength. Nylon stitching to lengthen seam life; close, firm stitches. High quality white rubber sole, heel. $5.95 [$20-50]

Above: **Four-eyelet U-throated Tie.** Perforated vamp and sides. $4.98 [$65-70] **Round-the-clock Slip-on,** "fringed" peplum, slim strap, stretchable throat. $4.98 [$45-50] **Vamp Ornament Shortens Your Foot.** Sling pump. $4.98 [$50-55]

Left: **Lacy Vamp Sandal.** $4.98 [$55-60]

Platform sandal. Wedge heel, Searosole. $4.98 [$65-70] **Trim Casual Tie.** Ease-afoot platform, wedge heel, Searosole. $4.98 [$45-50] **Four-eyelet Tie,** open vamp, comfort platform, wedge heel, Searosole. $4.98 [$65-70]

***Above:* High Vamp Suede Sling** pump, touched up with leather. $4.98 [$65-70] **Extra Support** in the built-up sides. $4.98 [$55-60]
***Right:* Saddle Strap Honey.** $5.49 [$45-50] **Carefree Sandal** perched on a flexible, "look-taller" platform. $5.49 [$50-55]

Suede Allied with Patent. Sandal charmer you'll see in classrooms, at coke sessions, under office desks,... $5.49 [$55-60] **Size-stealing Low V-throat** and foot-slimming diagonal strap. $5.49 [$60-65]

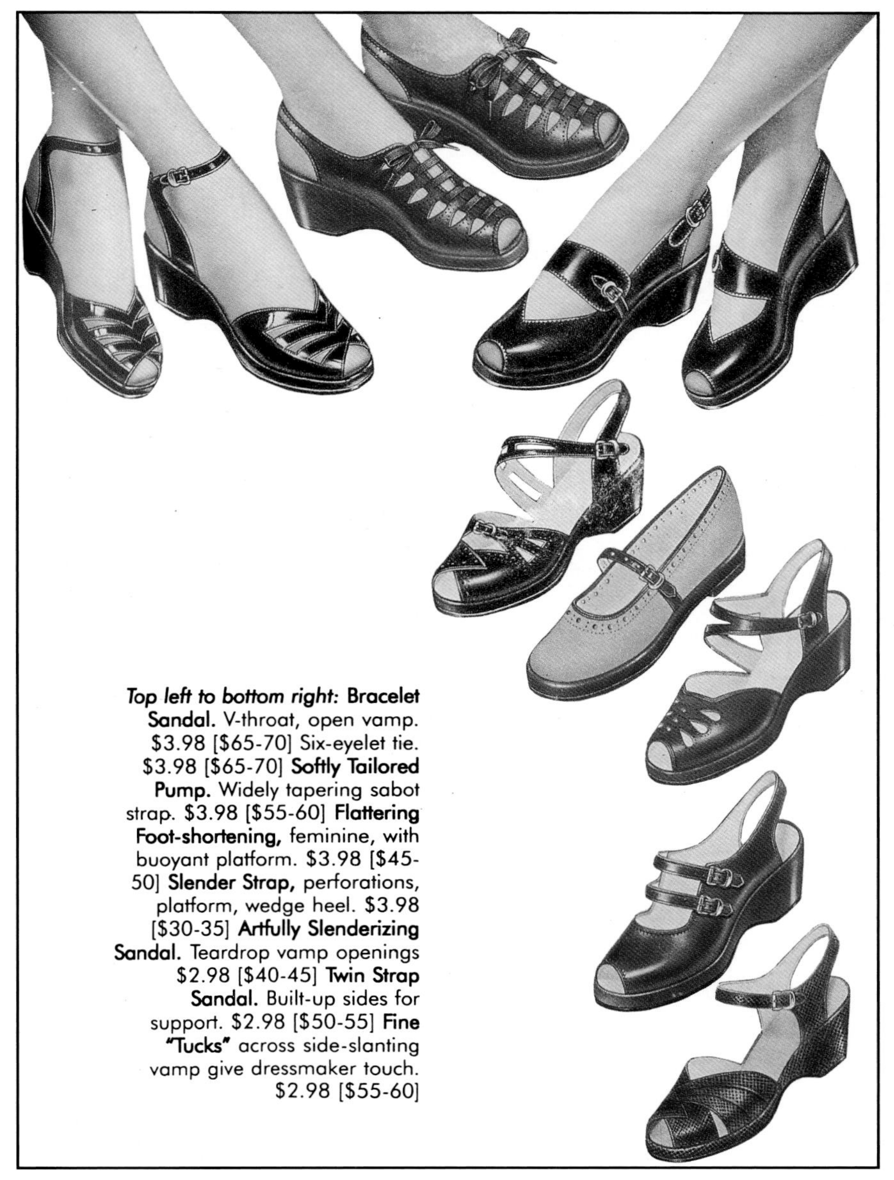

***Top left to bottom right:* Bracelet Sandal.** V-throat, open vamp. $3.98 [$65-70] Six-eyelet tie. $3.98 [$65-70] **Softly Tailored Pump.** Widely tapering sabot strap. $3.98 [$55-60] **Flattering Foot-shortening,** feminine, with buoyant platform. $3.98 [$45-50] **Slender Strap,** perforations, platform, wedge heel. $3.98 [$30-35] **Artfully Slenderizing Sandal.** Teardrop vamp openings $2.98 [$40-45] **Twin Strap Sandal.** Built-up sides for support. $2.98 [$50-55] **Fine "Tucks"** across side-slanting vamp give dressmaker touch. $2.98 [$55-60]

Top left to bottom right: **Smoothly Feminine Pumps.** One-piece leather cover over counter, 1-inch heel. $4.98 [$25-28] **Rounded-throat Flat.** Slim instep strap, .5-inch built-up heel. $2.79 [$30-35] **Tiny Perforations** around the collar, .5-inch heel. $3.98 [$25-28] **Baby Doll Flat.** For campus walking or waltzing, V-throat, widely spaced instep straps, .25-inch built-up heel. $3.98 [$30-35] **Feather-light Pumps.** Broad rounded foot-shortening toes, peaked vamp. $2.98 [$30-35] **Sparkling Ballerina** looks pancake flat, but there's a hidden wedge inside to make walking more comfortable. Rayon faille string bow. $2.98 [$30-35] **Light-hearted Flattie.** Low sweeping sides, V-throat, triple split strap. $4.45 [$40-42]

Note: Flat shoes were not quite as popular in the early 1950s as they are today. The styles pictured are particularly glamorous. Many women enjoyed wearing ballet flats, which are not shown in Sears catalog.

Top left to bottom right: **Companion for Fall Suits.** Wheeled extension edge leather sole, 2.25-inch heels. $8.90 [$50-55] **Slenderizing Envelope Vamp,** buckled vamp strap, 2-inch heel. $7.98 [$50-55] **Dressmaker Touch** scalloped throat line. $7.98 [$55-60] **Calfskin Pump,** square toe and heel, rope-stitched oval vamp ornament at throat, 2-inch heel. $7.98 [$55-60] **Spectators.** Calfskin perforated and pinked wall toe and back. $7.98 [$45-50] **Kerrybrooke Quality.** Slip-on; padded strap across U-throat, rope-stitched, ridged vamp, 2-inch heel. $8.90 [$45-50]

Above: Suede Sling Pump. Kidskin leaf design appliquéd on the side of the vamp, .25-inch kidskin covered platform, 3-inch full-breasted heel. $7.98 [$70-75] **Half 'n' Half** suede and kidskin, tiny tucks, V-line throat envelope vamp, full-breasted 2.75-inch heel. $6.98 [$65-70]
Left: Half Suede, Half Leather. Flattering D'Orsay line pump. $7.98 [$65-70]
Squared Throat and Heel. Trim sling pump, piped in contrasting color, .25-inch platform, 3-inch heel. $7.98 [$70-75]

Above: Shell-cut Vamp covered with strips, .25-inch platform, 3-inch heel. $5.95 [$65-70] **Dainty Sandal,** .25-inch platform, 3-inch heel. $3.98 [$65-70] **Hide 'n' Seek** toes. 3-inch heel. $3.85 [$75-80]
Right: Demure Baby Doll, side bow. $5.95 [$65-70]

Dainty Vamp Shirring, offside open toe, bracelet strap, 2.75-inch heel. $5.95 [$70-75] **Curving Leather Strips** cross the vamp, .25-inch platform, high 3-inch heel, leather sole. $5.95 [$70-75] **About the Prettiest Sandal** you'd ever find, 3-inch heel. $5.95 [$70-75]

Plain on One Side, tiny square cut openings on the other, V-throat line. $5.95 [$65-70] **Merry Sling Pump,** looped bow. 3-inch heels with .5-inch platform. $4.69 [$65-70] **Baby Doll Pump** ornamented with rayon faille side lacing. $4.69 [$55-60]

***Above:* Sweet Little Shell Pump** gives your foot a fragile, cherished look, rayon faille vamp strips and top binding. $8.90 [$60-65] **Wall-toe Pump.** Eggshell throat, .25-inch platform. $7.98 [$50-55] **D'Orsay Line Pump.** $7.98 [$50-55]
***Left:* Open Toe,** covered back; foot-revealing shell vamp. $8.90 [$65-70]
***Below left:* Calfskin Overlay,** scalloped at the side, combines with soft-as-moss suede, 3-inch heel. $7.98 [$55-60] **Gracefully Curved Throat,** .25-inch platform, 3-inch heel. $7.98 [$65-70] **Low-swinging Flattering D'Orsay Lines,** foot-slimming curved throat. $7.98 [$65-70]

Note: Low and high platform pumps from this period tend to bring the highest prices.

***Above:* Feminine Straps** on shell vamp, draped to look like bows, .25-inch platform, 2-inch heel. $5.95 [$70-75] **Black Beauties** with diamond-shape vamp openings, 2.25-inch heel. $7.98 [$70-75] **Flexible Kidskin,** smooth on one side, tiny openings on the other, 2.5-inch heel. $7.98 [$65-70]
***Right:* Shell Cut,** lattice vamp sandal, 2-inch heel. $8.90 [$55-60]
***Below right:* Rope-stitched Calf,** wide sabot strap, 2.25-inch heel. $7.98 [$65-70] **Provocative Sling Pump,** shell line vamp, a tiny peak at the throat; crisscross straps at the instep, 2.5-inch heel. $7.98 [$65-70] **Tiny Vamp Perforations,** white underlay. Broad strap, 1.5-inch heel. $6.95 [$55-60]

Note: The Nursery Rhyme Boots and Westernized Boots from the early 1950s, as well as the "young cowhands" styles and Roy Rogers cowboy boots crossover from the vintage clothing market into the collectibles market. For this reason, prices tend to be much higher for these than ordinary children's boots. Some items are so rare that collectors would pay an undetermined price to get them, so the price guide is less meaningful here than elsewhere in the book.

Nursery Rhyme Boots: Mary Had a Little Lamb, The Cow Jumped Over the Moon, Cat and the Fiddle, Humpty Dumpty, This Little Pig Went to Market, and Old Woman Who Lives in a Shoe. Felt. Padded sole. $2.27 per pair [$45-50]

All-rubber Cowboy-style boot. $3.49 [$55-60] **Roy Rogers** and Trigger on rubber boot. $3.79 [$85-115] **Hopalong Cassidy** and his autograph on a rubber overshoe boot. $3.89 [$85-115]

Toy Gun Totin' Holster. Felt boot. $1.98 [$55-60] **Imitation Suede** duded up with white trim, cotton fleece lined. $1.98 [$55-60]

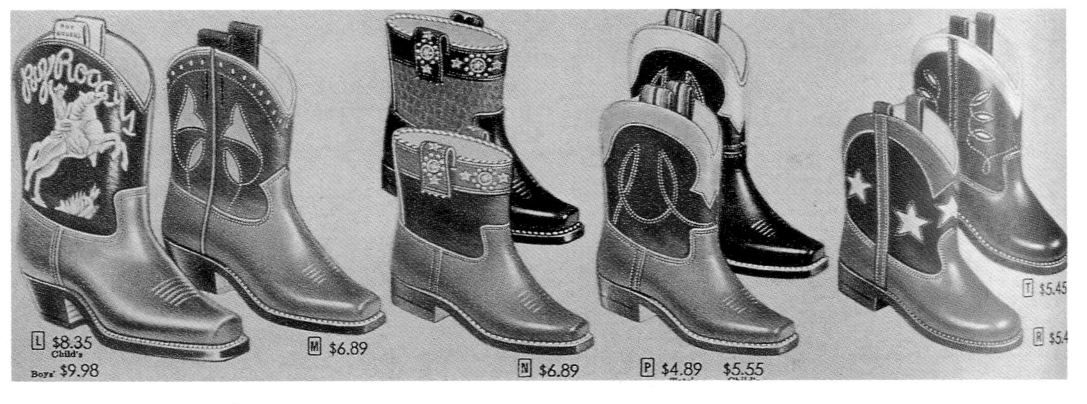

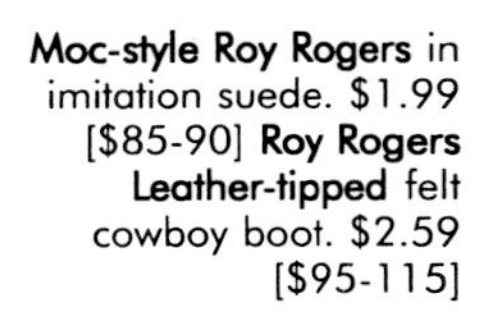

Just Like a Real Cowboy Boot. Durable leather, rubber heel. $3.29 [$55-60]

Moc-style Roy Rogers in imitation suede. $1.99 [$85-90] **Roy Rogers Leather-tipped** felt cowboy boot. $2.59 [$95-115]

***Above:* Roy Rogers, Trigger** embossed on leather boy's boot. $9.98 [$200-250] Cowboy Boots made in Texas, kidskin leg, leather vamp. $6.89 [$85-130] **"Wahoo" Boots.** Lower, more comfortable for children. Western jeweled collar, leather. $6.89 [$85-120] **Tots' and Child's** size boots. Kidskin leg, leather vamp. $5.55 [$80-110] **Lone Star** collar design. Leather uppers. $5.45 [$85-120] **Flat-heel Western Boots,** leather uppers. $5.45 [$85-120]
***Left:* Cowpoke Boots.** Embossed leg, heavy chain and saddle in nugget-colored (gold) metal, leather. $8.49 [$95-110] **Leather-lined Gaucho Boots,** ornamented leather cuff. $7.95 [$95-110]

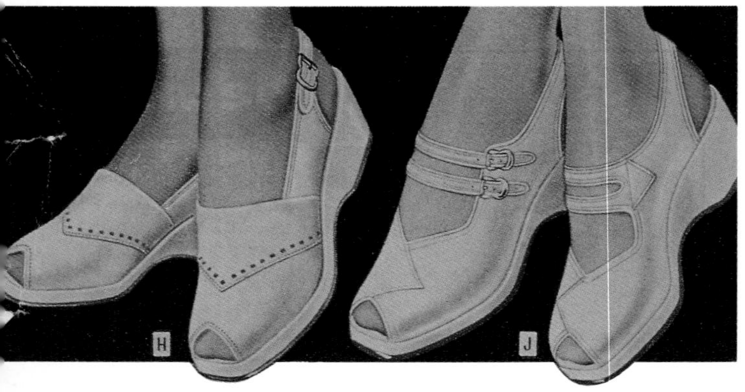

Heel-to-toe Foam Rubber cushions in sling pump, extra cushion supports your arch, 1.5-inch wedge heel. $4.57 [$60-65] **Twin-strap Sandal.** Platform, 1.5-inch wedge heel. $4.57 [$60-65]

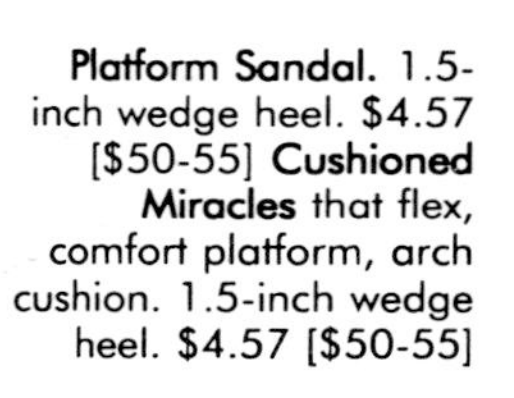

Platform Sandal. 1.5-inch wedge heel. $4.57 [$50-55] **Cushioned Miracles** that flex, comfort platform, arch cushion. 1.5-inch wedge heel. $4.57 [$50-55]

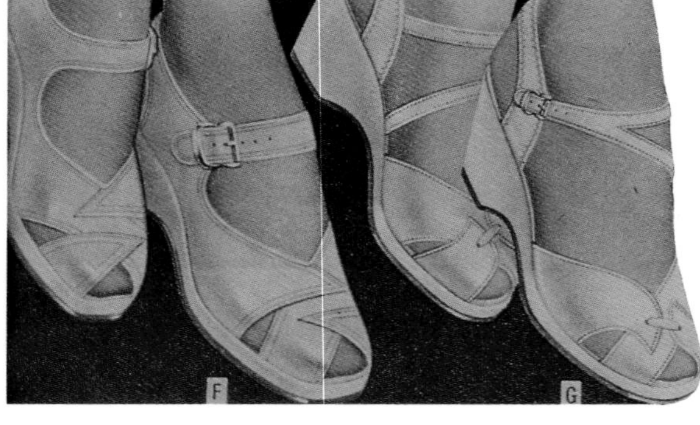

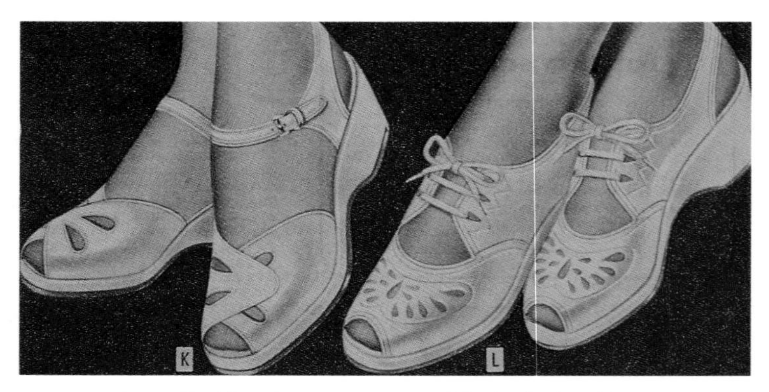

Inner Cushions in pretty strap sandal, shock-absorbing platform, 1.5-inch wedge heel. $4.57 [$60-65] **Clever Platform Tie,** 1.5-inch wedge heel. $4.57 [$70-75]

Above: **V-throat Pump.** $6.98 [$55-60] **Square-throat Calfskin Opera.** Perforated. $6.98 [$55-60] **One-strap Pump.** Flat as a pancake, .5-inch built-up heel. $3.98 [$30-35] **Exquisite Opera Pump.** Eggshell throat, kidskin. $5.98 [$55-60]
Lower left: **Wide Throat, Low-swinging D'Orsay** pump. $6.98 [$55-60] **Size-deceiving Pump** with pretty dress-maker details, .25-inch platform, 3-inch heel. $6.98 [$55-60]

Top left to Bottom right: **High 2.75-inch Wedge Heels.** Exotic sandals with Italian artistry in vamp and strap details. $4.98 [$55-65] **Romantic Riviera.** All raffia lined with smooth reinforcing material, .5-inch cork and rubber platform, 2-inch wedge heel. $3.98 [$45-50] **Woven Raffia Sandal,** ties bracelet style at your ankle, .5-inch cork and rubber platform, 1.25-inch wedge heel. $2.98 [$50-55] **Bright Cotton Frames** tapering vamp insert of contrasting color raffia. Raffia straps tie in a bow. $2.98 [$45-50]

Note: These wedge-heeled platform shoes were very popular in the early 1950s. My mother wore these with white ankle socks and printed cotton summer dresses, as she had during the late 1940s.

Left to right: **Pretty** with slacks and play clothes. Cotton tie, rubber platform. $2.49 [$25-30] **Fun-time Sandal.** Cotton braid, cork crepe rubber sole. $3.29 [$45-50] **Open Airy Vamp,** cork crepe rubber sole. [$40-45] **Multicolor** braid, cotton, rubber sole. $2.88 [$40-45]

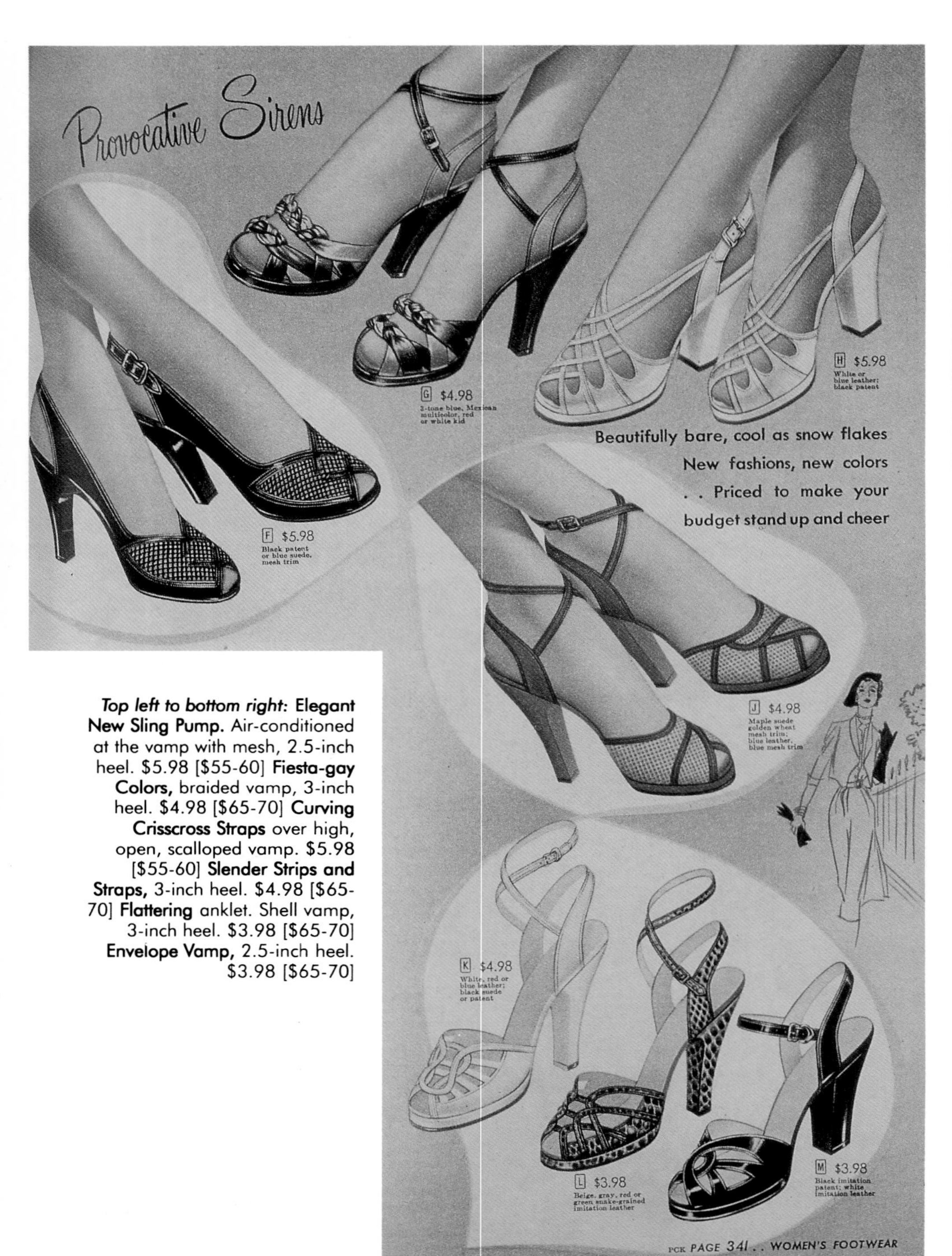

Top left to bottom right: **Elegant New Sling Pump.** Air-conditioned at the vamp with mesh, 2.5-inch heel. $5.98 [$55-60] **Fiesta-gay Colors,** braided vamp, 3-inch heel. $4.98 [$65-70] **Curving Crisscross Straps** over high, open, scalloped vamp. $5.98 [$55-60] **Slender Strips and Straps,** 3-inch heel. $4.98 [$65-70] **Flattering** anklet. Shell vamp, 3-inch heel. $3.98 [$65-70] **Envelope Vamp,** 2.5-inch heel. $3.98 [$65-70]

Suede with Kidskin. Low cut open vamp, little bow, 2.75-inch heel. $6.98 [$60-65] **Dance-loving Sling Pump.** Low cut vamp; wing shaped overlay, sparkling piping, 3-inch heel. $4.98 [$65-70] **Baby Doll Pump.** Low V-throat. $4.49 [$55-60] **Baby Doll Pump** ornamented with rayon faille side lacing. $4.49 [$55-60]

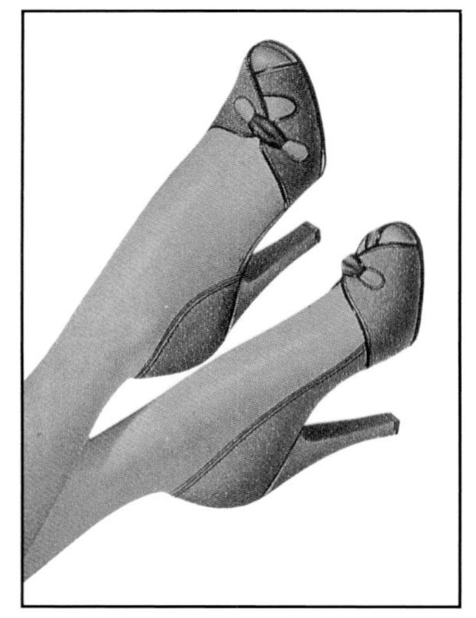

Festive Pump, open at one side, intriguing vamp design, 2.75-inch heel. $5.98 [$55-60]

D'Orsay line baby doll pump. $5.98 [$55-60]

Note: Once upon a time boys wore sneakers for gym class then threw them away at the end of the year! I do not remember my father ever wearing high-top sneakers during the 1950s. White low-cut canvas shoes were "required" for playing tennis in those days, and rubber platform shoes with a fabric mesh top were worn occasionally for leisure wear, but this was clearly "P.S.R." or pre-sneaker revolution! There is a market for the vintage sneakers because they really are so rare now. Prices vary tremendously for these early sneakers.

Army Duck Uppers. Molded-type non-marking rubber sole. Rubber toe guard and toe cap. $3.47 [$80-85]

Roy Rogers and Trigger imprinted on canvas uppers, gray rubber trim. $2.98 [$225-280]

Pro-type Styling. Cushion arch, duck uppers, ventilated openings, reinforced eyelets, molded rubber sole. $4.79 [$80-85]

Low-cut Oxford, duck uppers; cushion arch features, crepe rubber sole. $3.49 [$35-40]

Sure, Gentle Fit. Calfskin, 2-inch heel. $5.45 [$70-75] **Hand-turned Kidskin Slip-on.** 1.5-inch heel, rubber lift. $5.45 [$65-70]

Patent-sparkled Gypsy Tie. Five eyelets, 1.75-inch heel. $5.45 [$65-70] **Easy-going Wedge Tie.** Pliant kidskin upper, 1.25-inch wedge heel. $5.45 [$65-70] **Cork Cushions,** flexible kidskin, 1.5-inch heel. $5.45 [$55-60] **Twin-strap Envelope** vamp pump, 1.75-inch heel. $5.45 [$55-60]

Envelope Vamp Sandal. Soft platform, 1.5-inch wedge heel. $1.98 [$60-65] **Firm-feeling Bracelet Straps,** 1.5-inch wedge heel. $1.98 [$65-70] **Tiny-priced V-throat** casual, built up at the sides, broad rounded strap, 1.5-inch wedge heel. $1.98 [$65-70]

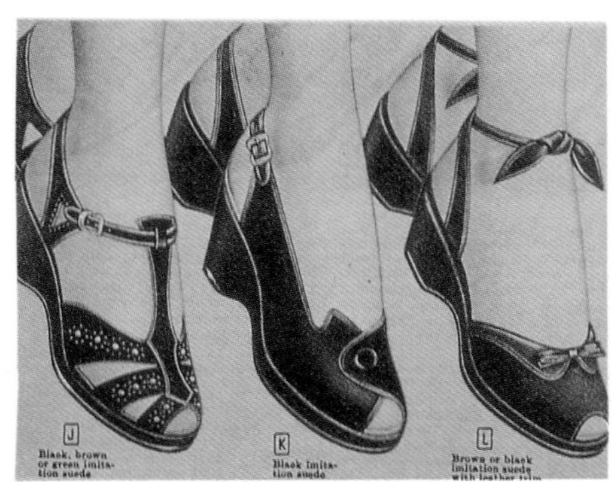

T-strap Sandal, easy-comfort platform, 4.5-inch wedge heel. $1.98 [$60-65] **Sling Pump.** Button-trimmed envelope vamp, adjustable strap high 2-inch wedge heel. $1.98 [$60-65] **Bracelet Strap Ties** kerchief fashion. Imitation suede and leather, 1.75-inch wedge heel. $1.98 [$65-70]

Note: Anything from the 1950s with leopard print is very popular with collectors. The leopard trimmed slippers and leopard print pairs shown in Sears catalog would be a great find now.

Warm Felt, gay leopard print trim. $1.59 [$60-65]

Broadly Cuffed Combat-style Boot. $3.98 [$30-35] **Low-cut for Wear** with rolled-up jeans, dandied up with flashy D-ring side pulls. $3.98 [$40-45] **High-up Leg Protection.** $3.98 [$30-35] **Young Bronco-busters' Range Rider** boot with a rough and ready cowboy embossed on leg. $3.98 [$95-110]

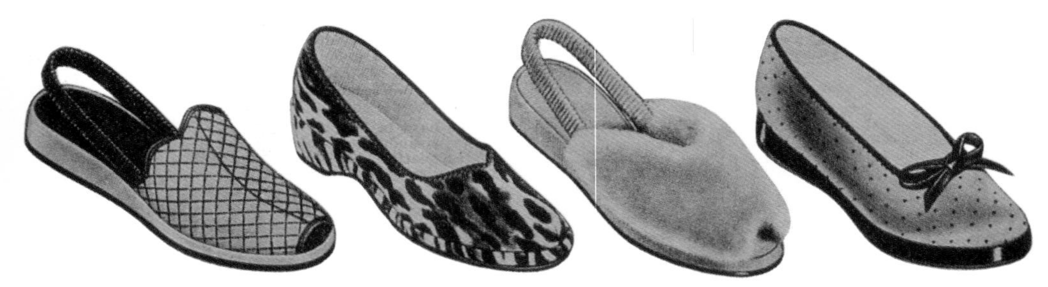

Rayon Satin, Scalloped Throat. $1.88 [$20-25] **Dramatic Leopard Print** in rayon and cotton. $1.88 [$65-70] **Deep Pile Shearling Scuff.** $2.98 [$25-30] **Radiant Rayon Satin,** polka dot trim. $1.79 [$30-35]

Suede and Leather Slippers, Canadian-made, featuring fur trim, hand beading, fluffy linings. $1.69-2.89 [$45-50]

Above: **Long Horn Design.** $4.98 [$85-90] **Lil' Dudes Western Boot.** $4.98 [$85-90] **Brawny Leather Field Boot.** $4.93 [$65-70] **Beltwel Rocket Boot.** $3.69 [$85-90] *Left:* **Roy Rogers** boots with fancy inlay, stitching. $7.65 [$200-250]

Gleaming Alligator-grained imitation leather sling pump, 2.5-inch heel. $3.98 [$85-90] **Demure Pump.** $3.98 [$55-60]

Flatteringly Simple Sandal. High 3-inch heel. $4.98 [$60-65] **Dressy Sling Pump.** Big bow makes foot look incredibly small, 3-inch heel. $4.98 [$65-70] **Shining Bracelet Sandal.** 38-inch heel. $3.98 [$65-70] **Baby Doll Pump.** Side vamp openings. $5.98 [$55-60]

Snake-grained imitation leather with imitation suede, 3-inch heel. $3.98 [$85-90] **Shell Vamp Sandal,** vamp openings, 2.5-inch heel. $5.98 [$75-80] **Suede Bracelet Sandal,** leaf design on vamp and heel, 1-inch platform, 3.5-inch heel. $5.98 [$95-100] **Bracelet Sandal.** Snow-flake pattern painted on vamp, flecked with rhinestones, 3-inch heel. $5.98 [$95-110]

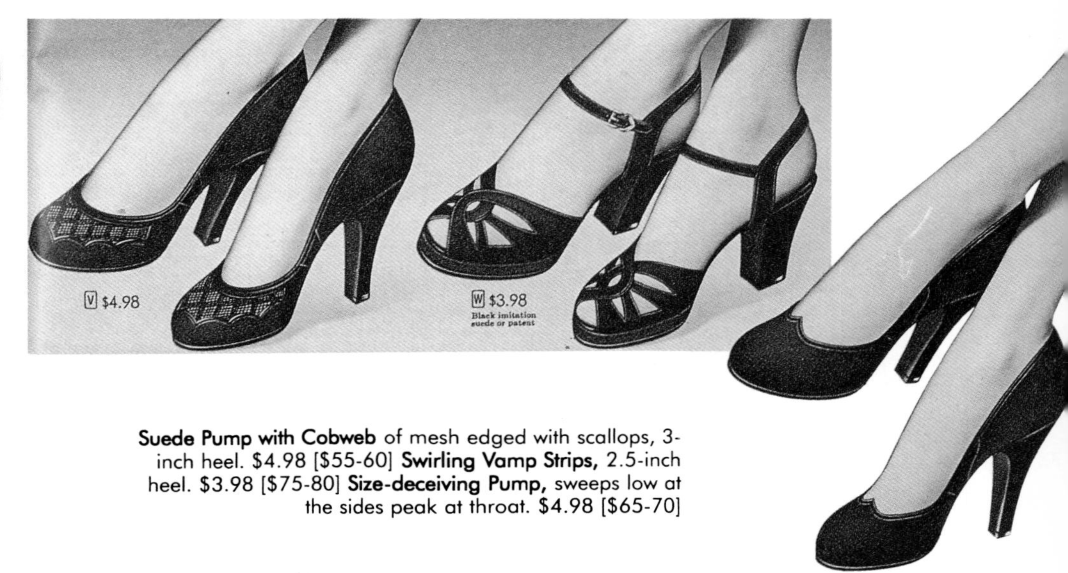

Suede Pump with Cobweb of mesh edged with scallops, 3-inch heel. $4.98 [$55-60] **Swirling Vamp Strips,** 2.5-inch heel. $3.98 [$75-80] **Size-deceiving Pump,** sweeps low at the sides peak at throat. $4.98 [$65-70]

Gypsy Oxford. 1.75-inch heel. $6.95 [$60-65] **Dress-up Open Vamp,** 2-inch heel. $6.95 [$60-65] **Stretchable Sides Slip-on,** 1.75-inch heel. $6.95 [$70-75]

Open-toe Pump. 2-inch heel. $6.95 [$55-60] **Dressy Oxfords.** Arch-supporting steel shank, 2-inch heel. $6.95 [$65-70] **Six-eyelet Tie.** 1.5-inch heel. $6.95 [$55-60]

Cotton Monk's Cloth Oxford. Cushiony cork and rubber sole wrapped in crepe-type rubber. $3.98 [$40-45]

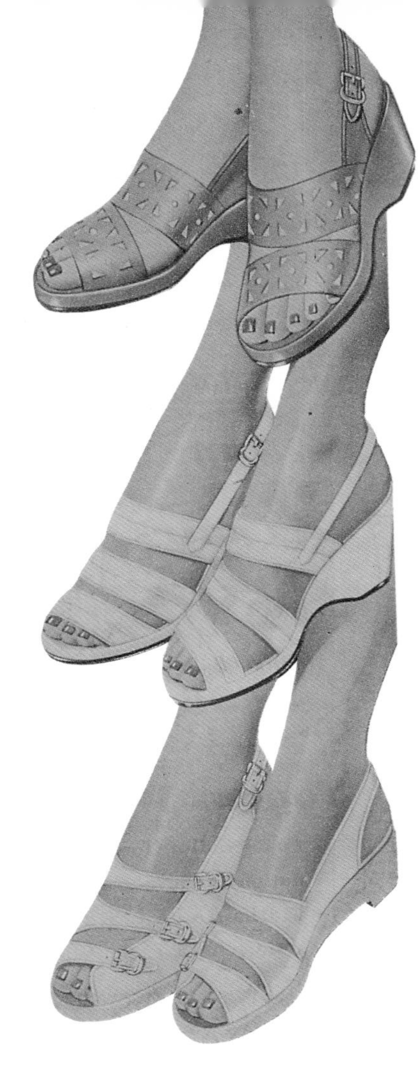

***Above:* Pleated Cuff,** finished with tiny buttons. High 2-inch wedge heel, $4.67 [$60-65]
***Left from top:* Two Broad,** opening-studded bands. 1.5-inch wedge heel. $3.98 [$45-50]
Banded Vamp Sandals, 2-inch wedge heel. $3.98 [$45-50]
Light-stepping Leather Sandals with bouncy crepe rubber platforms. $3.98 [$45-50]

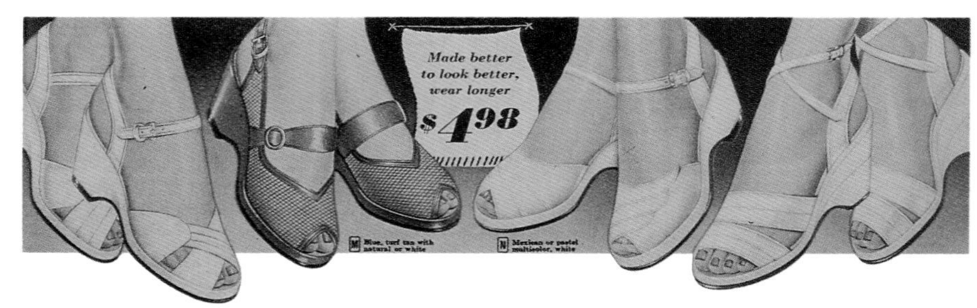

Softly Draped Vamp. 2-inch wedge heel. $4.98 [$45-50] **Nylon Mesh Sandals** trimmed with leather, 2-inch wedge heel, Searosole. $4.98 [$45-50] **Pastel or Mexican** multicolor, or white leather sandals. 1.5-inch wedge heel. $4.98 [$50-55] **Bracelet Sandal.** High 2-inch spool wedge heel. $4.98 [$60-65]

Best-selling U-wing tip leather oxford. Nylon mesh front and side invite cool air in. $8.85 [$75-80]

Gold Bond Slip-on styled for business wear, front-gored (with concealed elastic) for easy getting in and out. $8.85 [$85-90]

***Above:* Air-condition** your feet with a pair of ventilated Gold Bonds, priced to meet any budget. Sturdy deep brown leather, rich looking, supple; easy to shine. Long-wearing rubber sole and heel. Shape-retaining Goodyear welt construction. Brown. $5.95 [$65-70]
***Right:* Nylon Mesh Front** "fans" your feet while you walk. $5.95 [$65-70]

***Top row:* Rich Contrast,** light and dark, curved 2.75-inch heels. $6.98 [$55-60]
Band-box Fresh Sling Pump. Mesh and fine calf, scalloped collar, 3-inch heels. $5.98 [$50-55] **Nylon Mesh Wears So Well,** leather-bordered top line and toe, 3-plus-inch heel. $5.98 [$50-55]
***Right:* Crisply Tailored Spectators.** $4.98 [$60-65]
***Below right:* Cobweb Sheer Nylon Mesh,** calfskin bow, mudguard, 2-inch heel. $6.98 [$45-50]
***Below left:* Your Pet Spectator.** 2-inch heel. $6.98 [$60-65]

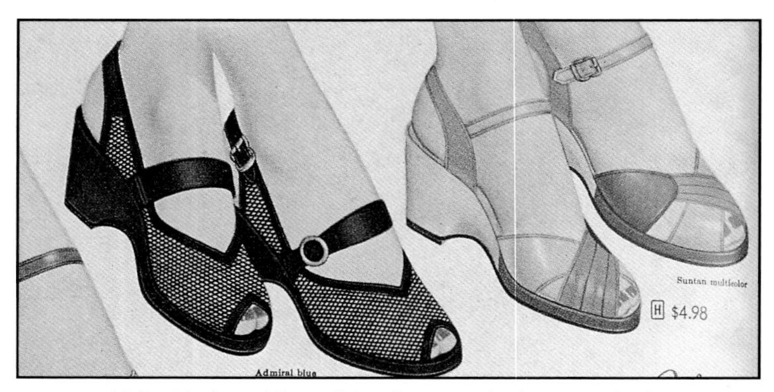

Nylon Mesh and Smooth Leather. Broad sabot strap, high 2-inch wedge heel. $4.98 [$40-45] **Softly Draped Vamp.** 2-inch wedge heel. $4.98 [$40-45]

Swirl Sandal. Crepe rubber platform, 1-inch wedge heel. $3.98 [$40-45] **Sling Pump.** Pleated cuff, finished with tiny buttons, 2-inch high wedge heel. $4.98 [$55-60] **Captivating Sandals.** 2-inch wedge heels. $3.98 [$50-55]

Top left to bottom right: **Exotic, Imported from Italy:** Raffia bands crisscross mesh vamp, 2.75-inch wedge heels. $4.98 [$55-60] **Airy Raffia Sandals,** cork and rubber platform, 2-inch wedge heel. $3.98 [$45-50] **Bright Cotton** frames tapering raffia vamp insert, 1.25-inch wedge heel. $2.98 [$50-55] **Woven Raffia Sandal,** ties bracelet style at ankle, cork and rubber platform, 1.25-inch wedge heel. $2.98 [$50-55] **Brief Barefoot Italian Design Sandal.** Raffia, cork and rubber platform, 1.25-inch wedge heel. $3.49 [$45-50]

Calf or Soft Suede Sandals. High 2-inch shaped wedge heel. $7.98 [$65-70] **Soft Slip-on Smoothie.** Crisscross braid; perforations, 1.5-inch wedge heel. $7.98 [$65-70]

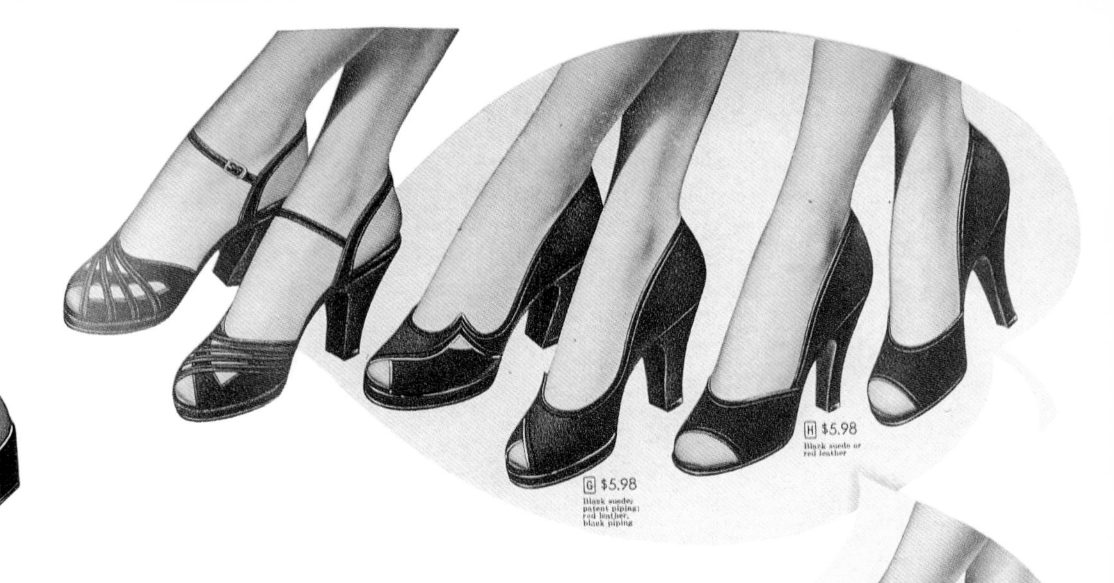

Note: White bucks joined saddle shoes in 1953 as a favorite among teenagers. Some teens associated the "white buck craze" with the singer Pat Boone, who had a very wholesome image. The shoes had to be cleaned with white powder and brushed off. I recall owning one pair of these shoes. They were tough to keep clean. You could only wear them with white cotton "bobby socks" and, although I didn't articulate it at the time, the image was a bit too wholesome for me. Elvis seemed a lot more exciting!

Goodyear Welt Saddles. $4.98 [$20-50] White Bucks—Hit Numbers! $4.98 [$20-50] **Red Rubber Soles.** White bucks with nylon seam stiching. $4.98 [$20-50] **Sturdily Crafted** of scuff-resistant leather. $3.98 [$20-50]

Above: **Network of Slim Strips** flared over vamp, 2.25-inch heel. $4.98 [$70-75] **Tiny Peak at the Side,** bright piping on envelope vamp, 2.5-inch heel. $5.98 [$70-75]
Right: **Wide Offside Toe Opening.** 2.5-inch heel. $5.98 [$55-60]
Below right: **Wide Side Opening.** Tapering swing strap, rounded throat, 2.5-inch heel. $5.98 [$65-70]
Below left: Crisscross Strap Sandal. Inter-locked open vamp 2-inch heel. $4.98 [$65-70]

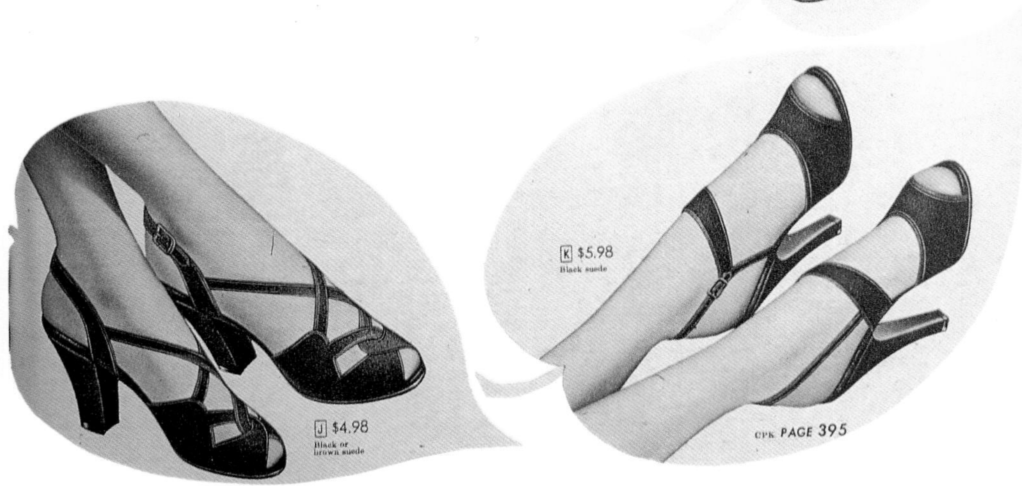

Note: The three styles of pumps pictured, show Sears' marketing concept of "good, better, best." Although you cannot detect the difference from the pictures, the descriptions detail the differences. Later Sears dropped the "good, better" lines and retained the "best." These pumps were very popular in 1953. I recall that my mother had black, brown, and navy in this exact style.

Good Value. Bewitching suede; eggshell throat, composition sole. 2 heel heights. $4.98 [$55-60] **Yes, It Costs $1 More.** Same style but made of finer leathers. Two widths, more sizes. $5.98 [$55-60]

Finest of the Three pumps featured here, looks better, feels better. Best quality suede or calfskin with leather soles. Extra pliable construction, Airfoam by Goodyear cushioned tread, perforated sock lining. $6.98 [$55-60]

Scalloped Mudguard of calf on flannel or suede. $6.98 [$55-60] **It's the Silhouette that Counts.** This one has flattering throat line. 2.5-inch new, more slender heels. $6.98 [$55-60]

Pilgrim Washfast Dress Socks for men. High-styled patterns alive with color. Guaranteed not to fade or stain in laundering. $.39-.98 pair [$8-15]

Note: Vintage cotton and rayon socks have become interesting to collectors, especially to women who enjoy wearing them with slacks or with sandals for summer. The most colorful socks, those with interesting patterns, bring the highest prices.

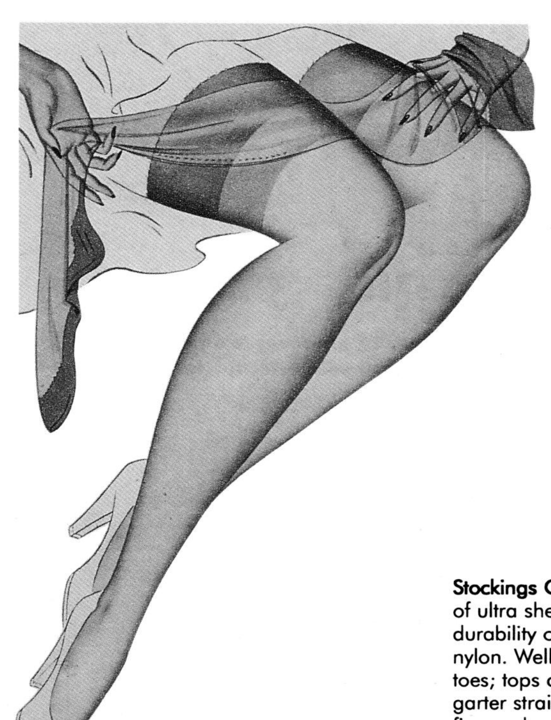

Note: During the early 1950s women wore nylons everyday. There really was no alternative. Seams were traditional, and as the 1950s progressed, "fashion marks" and seam detailing became very fashionable. There is no market for vintage 1950s nylon stockings, but the boxes these stockings came in are sometimes very interesting and great advertising collectibles.

Stockings Combine the Glamour of ultra sheer hose with the durability of service weights. All nylon. Well reinforced at heels and toes; tops are double to absorb garter strain. Seams in back are fine and even. $1.39 pair [NPA]

These All-nylon Mesh Beauties are guaranteed run proof. They look very sheer; actually they're made of heavier yarns than are plain knit hose of comparable weight. Flattering, even seams in the back. Reinforced heels, toes; double tops. $1.39 pair [NPA]

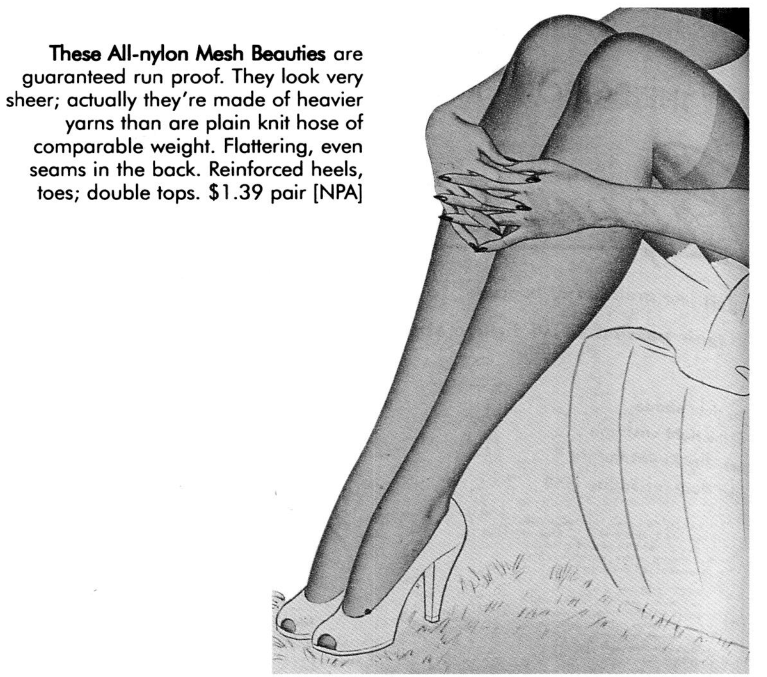

Heel is Rich, Deeper Tone of the same shade as the leg of the stocking. Designed for those who have yearned for ankle flattery without being "extreme." $1.44 pair [NPA]

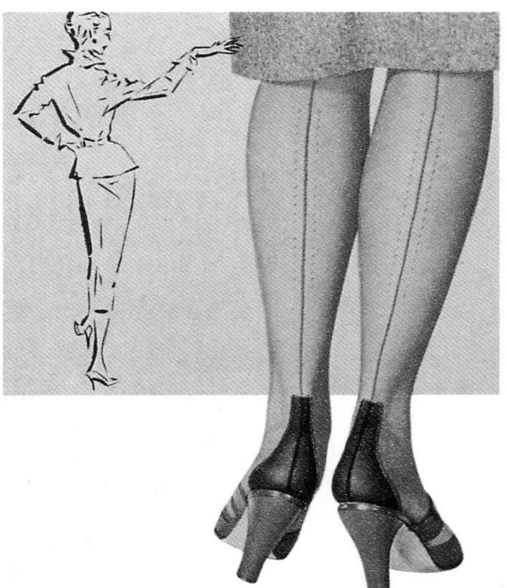

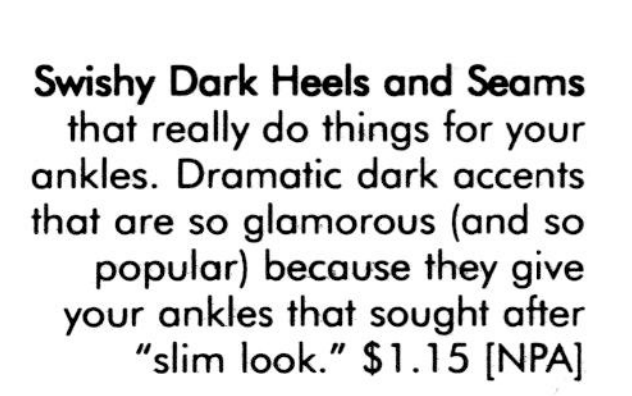

Swishy Dark Heels and Seams that really do things for your ankles. Dramatic dark accents that are so glamorous (and so popular) because they give your ankles that sought after "slim look." $1.15 [NPA]

Spring/Summer 1950

Fall/Winter 1953

Chapter 6

Nightwear and Underwear

Knee-deep Floral Design, tone-on-tone border overlaid all around. $5.59 [$65-70] **Queen of Hearts Robe.** Tufted hearts around whirl skirt, big pocket, double collar, inside ties. $4.98 [$65-70] **Diamond Checks** for smart contrast on big collar and extra large pocket. Inside ties. Belt tie from side seam. $4.79 [$60-65] **Baby Pinpoint Chenille** rows are finer, closer for smoother texture. Exquisite design around skirt. $6.22 [$65-70] All in copen blue, melon rose, aqua green.

Eighteen-inch Zipper closes cotton housecoat. Flattering collar, deep cuffs, pocket flaps. Blue, rose, lilac. $4.49 [$30-35] **Plisse Cotton Crepe** housecoat with eyelet embroidery at shoulders and clever tiers. Princess back. Navy blue, candy rose, aqua blue. $5.98 [$30-35] **Quilted Cotton Shortie** in a gay floral print. Turn-back cuffs, all-around belt. Gold ground, rose ground, green ground. $6.98 [$20-25] **Brushed Rayon Suede** hostess coat with fur-like leopard print collar, cuffs, back-buckled contour belt. Bright red or emerald green with leopard-like trim. $7.98 [$45-50]

Above: Two-tone Flowers and decorative overlays on pinpoint chenille. Flattering collar, sleeves, and skirt edges in contrasting tone. Pinpoint chenille in aqua green, melon rose, or copen blue. $4.98 [$40-45] ***Right:* Coachman-style Robe** in floral printed quilted cotton. Bright piping trim, gauntlet cuffs, big pocket. Copen blue, melon rose. $6.98 [$20-25] **Floral Print** quilted cotton robe curves its collar and big pocket sweetly. Print on blue or rose ground. $5.98 [$20-25]

Note: Rayon and gabardine men's robes from the early 1950s, especially satin-lined robes, are very collectible. Both men and women wear the elegant robes for leisure wear.

Rayon Jacquard Robe. Rayon satin shawl collar, cuffs, non-slip sash. $14.25 [$55-65] **Woven Stripe Robe.** Rayon with satin shawl collar, cuffs, non-slip sash, $15.98 [$55-65] Both in maroon or navy blue.**Woven Stripe Rayon Robe,** double-thick shawl collar, sash has non-slip feature, wide, shaped cuffs. $9.97 [$45-50] **Rayon Jacquard Robe.** Double-thick shawl collar. Non-slip sash. $8.97 [$45-50] Both in maroon or navy blue. **Polka Dot** trim. Rayon gabardine in solid color with polka dots on collar, cuffs, pockets, tops, and sash. Maroon, with white polka dot trim. Navy blue with white polka dot trim. $8.70 [$65-70] **Piping in Contrasting Color.** Rayon gabardine. Maroon with gray piping, navy blue with scarlet piping. $8.70 [$50-55]

Heavyweight Beacon Cotton Blanket Robe. $7.97 [$65-70] **Extra-heavyweight Beacon** robe. $9.35 [$50-55] **Woven Plaid** robe of 100 percent virgin wool flannel. Medium weight. $12.25 [$40-45] **Solid Color** robe of 100 percent virgin wool flannel. Medium weight. $10.98 [$40-45] **Notch-collar** style. 100 percent virgin wool flannel; medium weight. $12.60 [$40-45] All in maroon or blue.

Vat-dyed Pastels in broadcloth, gripper fasteners at waist, piping trim, tuck jacket inside trousers or wear outside. $3.49 [$30-35] **Broadcloth Type Cotton** in stripes and bold prints with white piping trim. $3.89 [$30-35]

Man-tailored Jamarette in glowing rayon satin. White piping trim. $3.98 [$35-40] **Mandarin-style Pajama** in rayon crepe. Contrasting piping. Trousers have elastic waist and buttoned side placket. $3.97 [$35-40] **Princess Style Jamarette** in cotton broadcloth. Sunny-hued prints, white rickrack trim. $2.99 [$30-35]

Above left: **Rayon Crepe Gown.** Lace trims neckline, cap sleeves. Matching lace midriff. Pink, yellow, light blue. $4.75 [$20-25] **Classic Gown** in rayon crepe. Charmode designed. Double fabric straps; slimming midriff.. Pink, yellow, light blue, pale green. $3.75 [$30-32]
Above right: **Chinese-style Shorty Gown.** Rayon crepe. Embroidered design of contrasting color. Piping on stand up collar, slit sides. Flamingo, aqua blue, yellow. $3.94 [$20-25] **Rayon Crepe.** Nylon and rayon lace. Pink, light blue, Nile green. $4.69 [$30-32] **Rayon Crepe.** Cotton lace straps and midriff inset. Pink, light blue. $3.50 [$30-32] **French-type Rayon Crepe.** Cotton lace trim. Self straps form V-back. Pink, blue. $2.39 [$30-32]

Above: **Dream Gown** with nylon net ruffle at neck and shoulders. Elastic cording gives pretty beaded effect. $3.68 [$20-22] **Lounger-sleeper.** Cotton lace trim at square yoke, cap sleeves. Covered elastic trouser waist. $3.94 [$35-40] **Cotton Lace Trim.** V-front and back, shirred midriff, gathered bodice. $3.68 [$30-32] **Lace at Peek-a-boo Midriff.** Matching lace forms straps and trims armholes. $2.59 [$30-32]
Above right: **Sweet Simplicity.** Surplice front drapes gracefully, elasticized midriff fits smoothly. $2.47 [$30-32] **Long-sleeve Gown.** Cotton lace trim. $3.99 [$35-40]

Rayon Bed Jacket for lounging or late reading. Charming over a gown. $2.35 [$12-15]

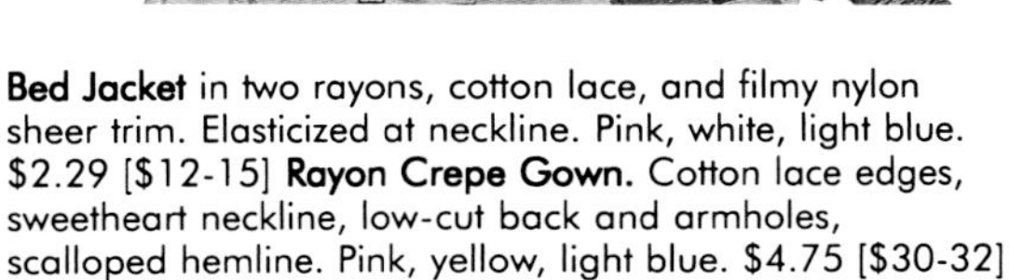

Bed Jacket in two rayons, cotton lace, and filmy nylon sheer trim. Elasticized at neckline. Pink, white, light blue. $2.29 [$12-15] **Rayon Crepe Gown.** Cotton lace edges, sweetheart neckline, low-cut back and armholes, scalloped hemline. Pink, yellow, light blue. $4.75 [$30-32]

Note: Chenille robes are extremely collectible, especially those with elaborate scroll work. The colorful floral prints for summer are glamorous for entertaining, as well as leisure wear.

Matching Cotton Jamarettes. Wrap-around Style Sleep Coat. $2.68 [$20-22] **Midriff Jamarette, Short Trousers.** $1.88 [$24-28]

Midriff Jamarette, Long Trousers. $2.28 [$28-32] In red, navy, gray print on white.

Note: There has always been a strong market for early 1950s nightgowns. Cut on the bias and trimmed with lace, these rayon nightgowns are sometimes worn as evening gowns.

***Above:* Cotton Plisse Crepe** housecoat with fresh white eyelet lace ruffles on wide rever collar, large pocket. Copen blue, rose. $4.98 [$45-50] **Polka Dot French-type** rayon crepe. Navy and white, bright red and white, aqua green and white. $3.98 [$40-45] **Floral Print** on cotton plisse crepe housecoat accented with white. Blue ground, rose ground. $4.98 [$45-50] ***Right:* Percale Zip-front Housecoat** in floral print. White eyelet ruffle, two handy pockets. Gray ground, fuchsia red print. $3.98 [$45-50] **Pinpoint Cotton Chenille Robe,** velvety, soft, comfortable to wear. Monotone floral overlay and loop scroll all around. Medium blue, melon rose, aqua green. $4.98 [$65-70]

Note: Collectors buy vintage 1950s full slips to wear under sheer dresses of the period, and with sheer blouses.

Above: Patented Eight-Gore Slip with cotton lace trim. $2.59 [$15-18] **Scalloped Nylon Ninon** sheer trim. $1.98 [$15-18] Rayon Crepe with nylon trim. $2.98 [$12-15] **Fine Woven Rayon Satin.** $2.39 [$15-18] ***Left:* Rayon Crepe Slips** with cotton lace trim. $2.69 [$15-18]

Trumpet Style Skirt of rayon crepe with cotton lace, scalloped appliqué on bodice. $2.94 [$15-18]

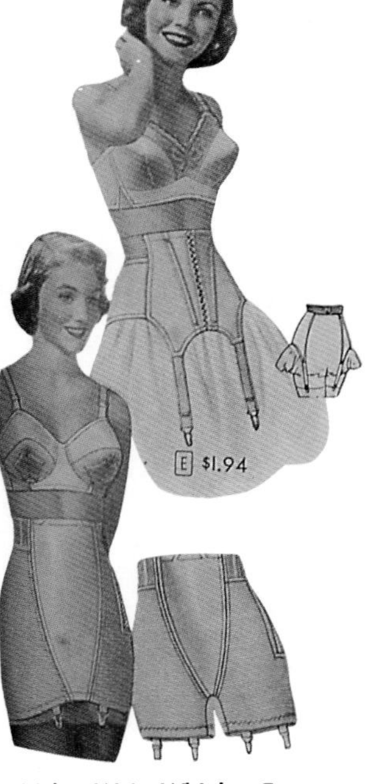

Nylon Waist Whittler. Front has fagot-stitching and two light bones for tummy control. Four elastic garters. $1.94 [NPA] **Nylon Tu-Way Control Girdles** with elastic sides and back. Short bones at top front and back. Zipper. $4.69 [NPA]

Boneless Nylon All-in-one. Nylon front for firm tummy control. Nylon back. $6.79 [NPA] **Tu-Way Control Styles** in three lengths. Double nylon front panel has two bones, satin elastic back. $6.79 [NPA]

Note: Foundation garments were an important part of a woman's wardrobe in the early 1950s. Although there is not a significant market for these items, lingerie collectors and designers are looking for a variety of styles.

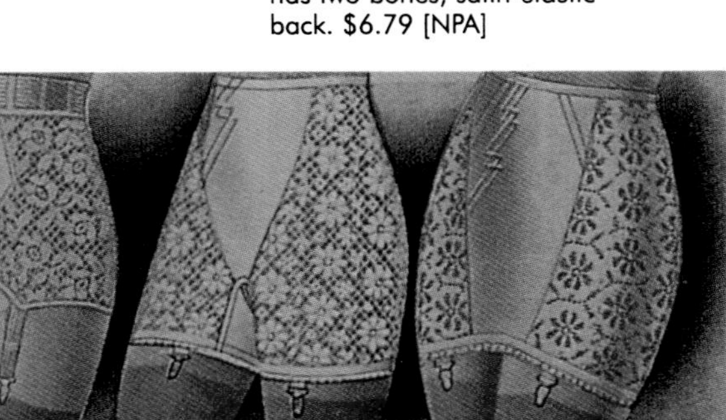

Lace Elastic Waist Whittler. Rayon satin elastic front panel. Coiled wire boning at sides and short bone at front. $3.94 [NPA] **Lace Elastic Briefs. Cotton,** rayon, and nylon elastic. 2 short bones at top, four garters. $3.39 [NPA]

Three-way Duster in baby pinpoint chenille. Wear it belted, half-belted, or pyramid style. Multicolor design on pockets. Double Peter Pan collar. White, aqua green, melon rose. $4.98 [$60-65] **Chenille Robe** with double revers, rich design around skirt. Copen blue, melon rose, aqua green. $4.98 [$65-70]

Flattering Rounded Yoke and the deep-fitted cuffs are all smartly quilted. Navy blue, aqua green, melon rose. $6.98 [$35-40] **Glitter Duster** in at jaunty length. Rayon crepe print sparkles with gold-color tracery. Gold-color braid on mandarin collar. Black, melon rose, or aqua green. $3.98 [$25-30] **Hand-painted Look** acetate jersey housecoat with silk-screen floral print on rounded yoke and big pocket. Aqua green, melon rose, navy blue. $7.98 [$45-50] **Polka Dots** in silky rayon crepe. Twin-button band. Navy blue, raspberry, and aqua green. $3.98 [$45-50]

Note: Many of the fashions for teenagers and children during the early 1950s are basically smaller versions of the adult styles, as in the case of the chenille and quilted robes.

Cotton Zip-front Housecoat with autumn leaf print, white piping. $2.98 [$25-30] **Quilt-trimmed Plisse Cotton** double-breasted coachman style with winged Dolman sleeves. $4.98 [$30-55] **Cotton Plisse Crepe** floral print with eyelet embroidery. $3.98 [$40-45] Flannel Duster with gold-color piping, small collar, deep raglan sleeves, turn-back cuffs. $5.98 [$20-25]

Above right: **Cotton Chenille** in two-tone overlay pattern. Aqua blue, melon rose, copen blue. $2.98 [$40-45] **Belted, Half Belted, or Duster** style. Scalloped piping, double pocket. Light blue, bright red, yellow. $4.98 [$20-25]
Above: **Pinwale Corduroy Duster** with mandarin collar, gold-color buttons. $6.98 [$15-18] **Duster with Cute Peter Pan Collar,** smart back flare, gold-color buttons. $5.98 [$12-15] Both in melon pink, aqua green, copen blue.
Right: **Floral Duster Style** with mandarin neck, deep raglan sleeves. Navy, red, green. $5.98 [$15-20]
Flattering Rever Collar. Sleeves with turn-back cuffs, tone-on-tone overlay design on skirt. Aqua green, melon rose, copen blue. $3.98 [$40-45]

Acetate Crepe Pajamas: Mandarin Style Pajama. Matching scuffs included, long-length jacket, black trousers. Aqua blue, rose print. $3.89 [$40-45] **Mandarin Style** with screen print trim. Pink coat with wine trousers. Lime green coat with navy trousers. $3.39 [$40-45] **Satin Jamarette Gripper.** Fasteners at trouser waist. Wine red, bright blue. $3.98 [$45-50] **Tailored Pajama** in two-tone style with white piping trim. White, peacock blue. $2.98 [$35-40]

Charmode Jamarettes. Cotton broadcloth in assorted prints. $3.29 [$25-30]

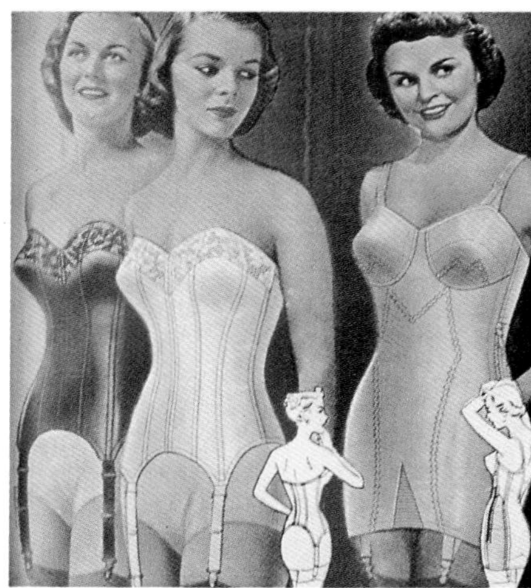

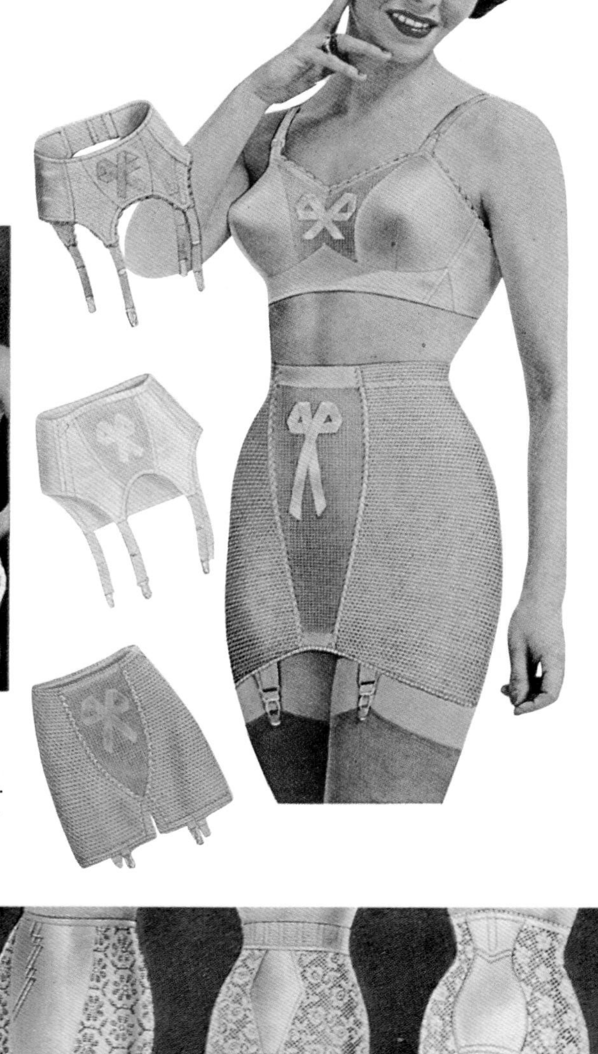

***Above:* Figure Charmer.** Satin: $4.49 [$25-30] Nylon: $5.39 [$25-30] **Boneless All-in-one.** $6.79 [$20-25] ***Right:* Bra.** Nylon. $4.94 [NPA] Girdle. $3.94 [NPA] **Regular-style Garter Belt.** $1.49 [NPA] **Apron-back Garter Belt.** $1.94 [NPA]

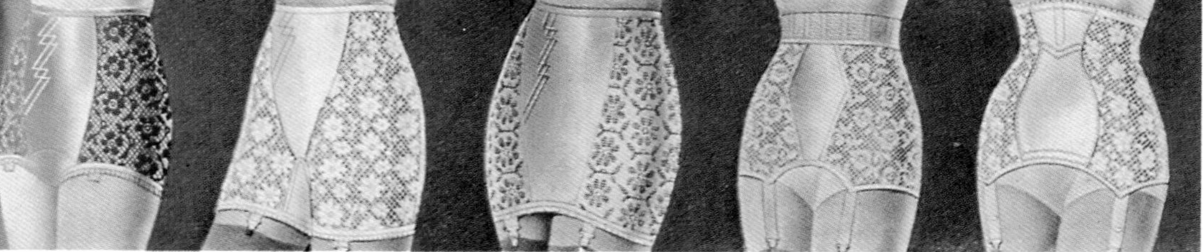

Assorted Girdles. In white, pink, blue, black, and flesh pink. $2.94-4.29 [NPA]

Charmode Jamarettes in cool, absorbent cotton plisse. Washable. Fine stitched, strong buttonholes. U-shaped crotch for extra ease, unhampered freedom: **Mandarin Styles.** Multicolor novelty print on white. Pastels. $3.49 [$20-25] **Man-tailored styles.** Fringed belt. Pink and blue print on white. Pastels with white piping trim. $3.49 [$20-25]

Two-piece Midriff Style Jamarettes: Short Trouser Styles. Multicolor print on white. Nile green, maize. $2.49 [$25-30] **Long Trouser Styles.** Multicolor print on white. Nile green, maize. $2.98 [$25-30] **Shorty Gowns.** White eyelet trim. Major seams are double-needle stitched. Washable. Nile green $2.45 [$25-30], pink $2.29 [$12-15]

Mother Hubbard Gown. Double fabric front and back yoke. Pretty pink rosebud print on white. $2.48 [$12-15] **Cotton Plisse Gown.** Covered plastic shirring at neckline, puffed sleeves. Cotton eyelet ruffle trim. Pink, light blue. $1.98 [$20-24] **Cotton Plisse Gown** with cotton eyelet trim. Prettily ruffled cap sleeves. High back. Self belt. Multicolor print on white. $2.48 [$24-28]

***Above:* Silky Soft Crepe of French-type** acetate crepe, quilted at collar, cuffs, and pocket. $3.98 [$30-35] **Polka Dot French-type** rayon crepe in double-buttoned style. Navy blue, raspberry, aqua green. $3.98 [$40-45] **No-iron Plisse,** floral on colored ground. Eyelet ruffles. Gray, melon rose, aqua blue. $3.98 [$50-55] **Zipper Front** housecoat with flowers on cotton. Ruffled front yoke. Red, navy blue. $2.98 [$50-55] ***Left:* Plisse with Hand-painted Look.** Modern motif on bodice and skirt. $3.98 [$55-60]

"Harem Trellis" Ensemble. Inspired by the Orient, styled by top designers, suggested by Sears for its graceful motif, thrifty price. Bra: $1.49 [$18-20] **Regular-style** garter belts: $1.49 [$18-20] **Long Apron-back Garter Girdle:** $1.94 [$18-20]

Chapter 7

Coats, Suits, and Jackets

Note: The contemporary fashion market offers little excitement in women's outerwear. Coats tend to be plain, almost drab, without cuffs or detailing, and with little or no interest to the backs. The coats of the early 1950s offer a myriad of styles, wide cuffs, covered buttons, great swing, and detailing. These coats are big sellers because collectors find they pay a fraction of the price of a new coat for a more beautiful, flattering style than what is being offered on the contemporary market.

Wing-scalloped Yoke enhances a cascading back, parts to show triple set closing. Wool with rayon satin lining. Navy blue, medium gray, emerald green. $19.50 [$95-110] **Terrific Over Suits.** Softly flared back, generous front overlap, big patch pockets. Wool, lined with rayon satin. Navy blue, medium gray, emerald green. $16.98 [$85-90] **Paris-inspired Fashion.** Wear with the triple pleats bloused softly or swinging free. Wool with rayon satin lining. Navy blue, medium gray, light brown. $18.98 [$95-110]

***Above:* Two-way Raincoats:** Full flare below a deep back yoke. Front has a four-button closing, big patch pockets. Medium gray, bright red, emerald. $8.98 [$50-55] **Fine Fabric Falls** with a rich drape in the graceful flare back. Medium gray, bright red, black. $8.98 [$55-60] **Candy Stripe Trim** of water-repellent rayon taffeta trims deep pockets and cuffs on this neat double-breasted two-way. Medium gray, emerald green, navy blue. $8.98 [$55-60] ***Left:* Water-repellent Rayon Gabardine** with rich water-repellent lining. Lined detachable hood. Medium gray, emerald green, black. $12.50 [$55-60]

Above: Leisure Jacket of Rayon Gabardine. Back cut full with side gussets, three-piece belt; detachable front sections, saddlebag pockets. $14.95 [$75-80] **Sash-belt Style** designed and made in California, headquarters for smart styling. Rayon gabardine $10.95 [$85-90] **Four-patch Pocket Style** in sheen gabardine, 40/60 virgin wool worsted/rayon. Padded shoulders for authentic "he-man" effect. $14.50 [$75-80]

Easy, Casual Fit and husky, athletic lines. All-rayon gabardine with padded shoulders, patch pockets, and durable construction. $8.75 [$75-80] **Rayon Fabric in Marbleized Tones.** Tailored in the California manner with shirred yoke in front and back, front waist tucks, stitched-down belt, and saddlebag pockets. Masculine one-button style. $12.95 [$75-80]

Note: Men who are looking for great style collect the gabardine jackets from the early 1950s. There's nothing else like them.

Boyville's Finest Two-ply Gabardine. Zipper front, belted sport back. $6.49 [$95-110]

Above: **Easy Shoulder, Wide Armhole,** side fullness, and stand-up collar characterize the best in pyramid styling. Dark green, dark gray, maroon wine, taupe. $19.98 [$95-110] **Classic, Flare-back Coat** treated to double tabs at pockets. Dark green, dark gray, maroon wine, taupe. $19.98 [$95-115] **Pyramid Look** in a restrained manner with softened shoulder, gathered cuffs, and gentle fullness. $19.98 [$95-110]
Left: **Skillful Tucking** and bracket-shaped flaps in front, back yoke, center pleat, and more tucks to cover every fashion focal point. $19.98 [$95-110]

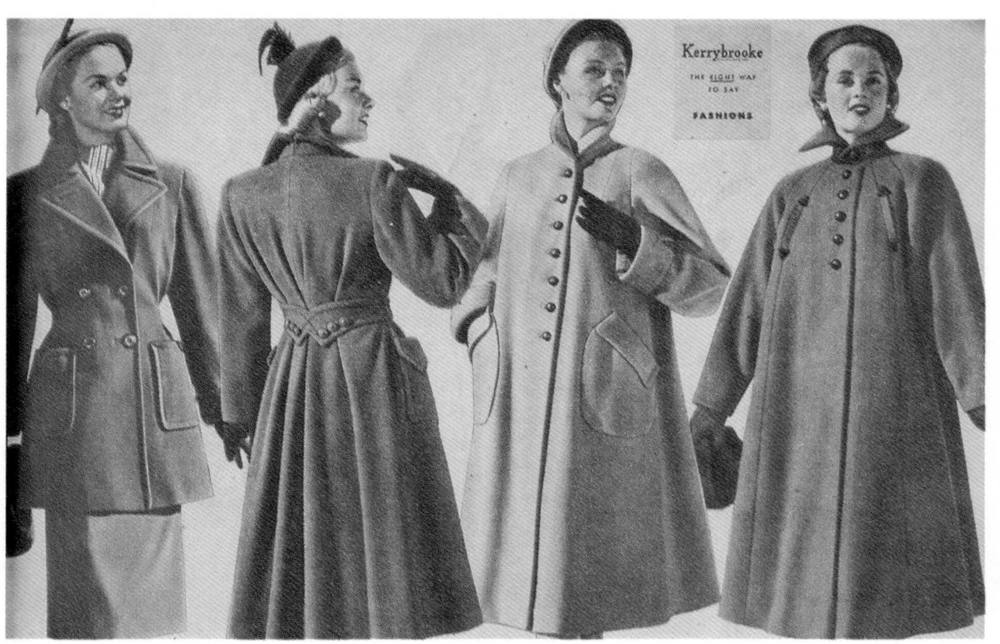

Three-quarter Length Coat with inverted box pleat in back. Rayon satin lined; warmly interlined. Bright red, rum, gold, navy blue. $23.98 [$55-65] **Wrap-around** with figure-enhancing back interest that begins with button-trimmed contour belt and continues with rippling fullness to the hemline. Convertible stand-up collar. $27.98 [$95-120] **Easy-fitting, Conservative Interpretation** of the new pyramid without that noticeable fullness. $27.98 [$85-115] **Full, Sweeping, Billowy Pyramid.** Underside of stand-up collar and arrow points are rayon velvet. $29.98 [$85-110]

Year-round Zip-outs in sheen gabardine: **Pyramid Silhouette** for style, plus a pliable zip lining of pure wool for warmth. $27.50 [$85-115] **Rayon Sheen Gabardine** for shorter women. In dark gray, dark green, or maroon wine. $29.50 [$95-130] **Dressy Accent** with convertible collar in gleaming rayon velvet. Dark green, dark gray, maroon wine, black. $27.50 [$85-115]

Utility Coat for driving, hiking, keeps legs and feet free Mouton-dyed lamb collar, alpaca pile lining, quilted rayon lined sleeves. Half belt and action box pleat in back. $26.75 [$55-60] **Queen of Storm Coats,** belted with full collar of mouton-dyed lamb, alpaca pile lining. Elastic shirring in back for waist-hugging fit. $28.75 [$60-65] **Practically Necessary** for rugged, blustery weather. Button up mouton-dyed lamb collar and matching belt, alpaca pile lining. $29.75 [$45-50]

Wrap-around with button-trimmed contour belt and a cascade of fullness to the hemline. Wide, stand-up collar. All-wool suede cloth. $7.98 [$85-95] **Modified Pyramid** without that noticeable fullness. Ripple-weave suede cloth lined with matching rayon satin, warmly interlined. $27.98 [$85-95] **Arrowhead and Velvet** trimmed full pyramid in ripple-weave all wool suede. Bright red, rum, gold, navy blue. $29.98 [$85-95]

Wool Sheen Gabardine Coats: Scalloped Yoke Classic. $44.50 [$95-120] **Zip-out Shawl-collar Pyramid.** $44.50 [$95-110]

Winter Storm Coat lined with warm wool alpaca pile, wide convertible collar. Two pockets, self belt, pleated back, bone buttons. $29.98 [$55-60] **Back Belted Wool Fleece** lined with rayon and cotton flannel. Smartly shaped back belt with pleats above and below. Double breasted, two novelty patch pockets. $24.98 [$40-45]

Collegiate Corduroy Fashion has tab collar that buttons up to show a plaid rayon facing, like the jackets worn by campus men. Pinwale corduroy lined with matching rayon. Leather buttons. $11.98 [$35-40] **Wool Flannel Casual.** Pointed collar with metal button, two pockets with metal buttons and buttonholes. $9.98 [$50-55] **Wool Plaid Tweed** has homespun look with lines of contrast color on a deep menswear gray ground. $14.89 [$35-40]

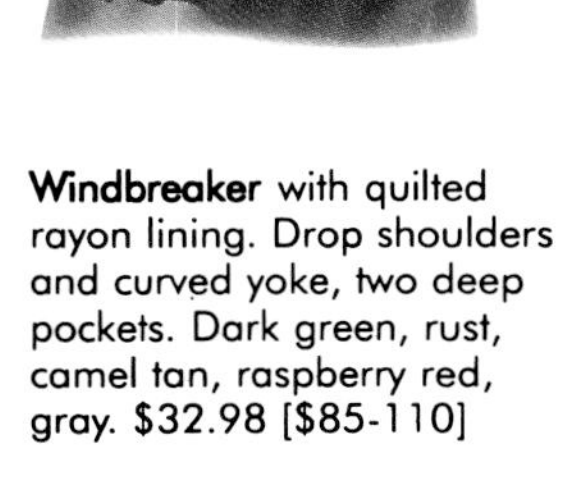

Windbreaker with quilted rayon lining. Drop shoulders and curved yoke, two deep pockets. Dark green, rust, camel tan, raspberry red, gray. $32.98 [$85-110]

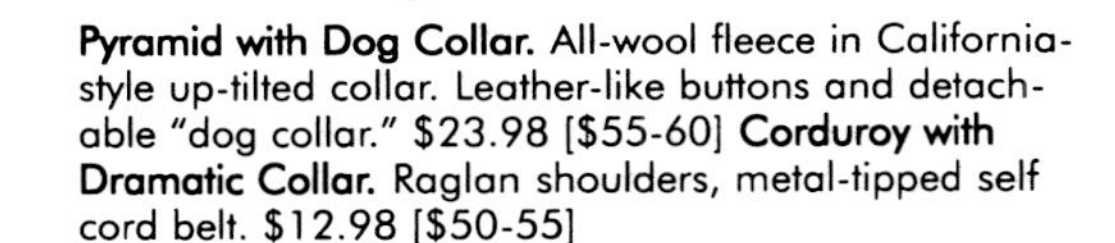

Pyramid with Dog Collar. All-wool fleece in California-style up-tilted collar. Leather-like buttons and detachable "dog collar." $23.98 [$55-60] **Corduroy with Dramatic Collar.** Raglan shoulders, metal-tipped self cord belt. $12.98 [$50-55]

Water-repellent Cotton Poplin with zipper front, two-way saddlebag pockets, adjustable waist tabs. Natural tan, light gray. $5.98 [$55-60] **Pinwale Corduroy Jacket** with becoming flange pleat shoulders, self belt with covered buckle, two big pockets with flaps. Dark green, raspberry wine, deep rust. $8.98 [$40-45] **Bright Embroidered Crest** on the upper pocket of sheen rayon gabardine jacket. Novelty metal buttons. Three-patch pockets. $12.89 [$40-45]

Yoke Classic Zip-out. Sheen gabardine with removable lining. $42.50 [$95-120] **Gleaming Touch of Velvet** on the underside of the convertible collar. Sheen gabardine. $44.50 [$95-115]

Note: In the early 1950s young girls wore "snow suits," which had flannel-lined "leggings" or snow pants and jackets, sometimes with hoods. Often these suits were worn over the girl's dress. I can remember wrapping my dress around me and pulling up the "leggings." There was always a bulge around the middle. The "leggings" were very warm, and the jackets were "pinched" at the waist, sometimes belted, and usually zippered. Little girls dreamed that some day they would get out of "leggings." It was a right of passage in those days.

Pyramid Coat. Easy-on shoulders, cotton velvet trim, mandarin collar, four-gore back flare, in 100 percent reprocessed wool coating. $17.95 [$75-80] **Shirtwaist Pyramid.** Tricky stand-up collar, full cuffed sleeves, pyramid back flare. $18.95 [$75-80] **The Peplum Coat** dressed up with velvet, petal collar, of 100 percent reprocessed wool. $19.95 [$70-80] **Popular Swing Coat** with scalloped yoke, gored for back flare, of 100 percent reprocessed wool. $16.95 [$70-80]

Hooded All-wool Suit. Zip-front jacket, hood, contrast trim, cotton kasha lined. $10.95 [$40-45] **Fur-Collared Suit** with Alpaca lining, rayon twill jacket, mouton Lamb collar, reprocessed wool slacks. $14.95 [$40-45] **Plaid 'n' Plain Suit** of wool with zip-closing, novelty metal buckle. $13.95 [$50-55] **In Two Fabrics.** Zip-front shirred jacket with contrasting collar, tabs, pockets, and snug-knit wristlets. $12.95 [$40-45] **Our Best Quality** water-repellent suit of 20/80 nylon/rayon plus mouton lamb collar and wool-quilted-to-rayon lining. $16.95 [$40-45]

Bomber Jacket of rayon and nylon gabardine. $9.95 [$65-70] **Stylish Surcoat Length** in rayon and nylon gabardine. $11.70 [$65-70]

***Upper row:* Wool Plaid.** Practical shirt-style jacket with coat-style sleeves. $5.95 [$40-45] **Gabardine Surcoat.** Water repellent rayon gabardine with wool lining. $9.85 [$65-85] **Pinwale Corduroy** with quilted rayon lining. $9.20 [$40-45]
***Lower row:* Zipper Front Corduroy** with cotton plaid lining. $6.24 [$40-45] **Rayon Gabardine** with rayon lining. $7.45 [$50-55] **Wool Plaid Jacket** with sueded cotton lining. $7.50 [$45-50]

Fashion Tailored of all rayon Zelan-treated gabardine. Body-and-sleeve zip-lining of rayon quilted-to-wool. $22.50 [$65-70] **Sheen Gabardine** with inverted pleating on pockets, shirred yoke back; padded shoulders. $14.95 [$65-70]

All-rayon Gabardine with shirred yoke, padded shoulders, saddlebag pockets, removable tie belt. Full rayon lining. $11.95 [$65-70] **Serviceable Garment** of virgin wool. Solid-color model gives you a lot for your money. $9.75 [$40-45]

Fashion Tailored rayon gabardine leisure jacket. $17.95 [$60-65] **Slacks:** Fine gabardine of 80/20 rayon/nylon. $6.60 [$50-55] **Pinwale Corduroy** leisure jacket with shirred yoke, padded shoulders, saddle-bag pockets, removable tie belt. $13.50 [$50-55] **Slacks:** All-rayon gabardine. $5.85 [$50-55]

Corduroy 31-inch Surcoat. Fully lined with rayon quilted to 6-ounce reprocessed wool. Zip front, padded shoulders, interlined collar. $12.75 [$45-60]

Newest Racer Model. Tough, shiny satin-back twill motorcycle jacket of rayon and cotton, detachable lambskin collar, lined rayon and wool, 7-inch map pocket, glove pockets, and gusset cuffs to fit tight under gloves. Bi-swing action back for greater arm and shoulder freedom. $14.50 [$175-250]

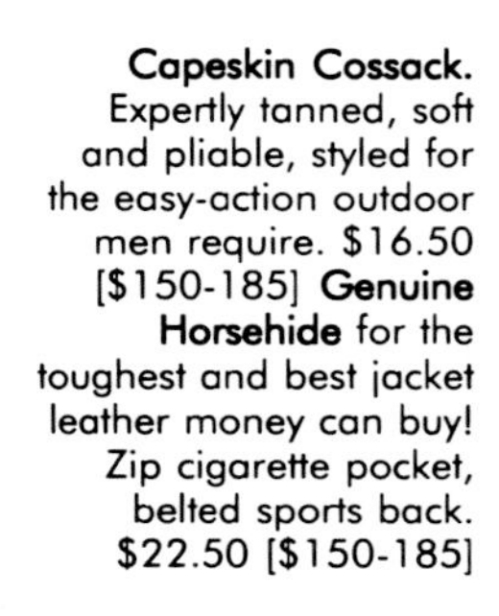

Capeskin Cossack. Expertly tanned, soft and pliable, styled for the easy-action outdoor men require. $16.50 [$150-185] **Genuine Horsehide** for the toughest and best jacket leather money can buy! Zip cigarette pocket, belted sports back. $22.50 [$150-185]

Horsehide Blouse. Lined with rayon quilted to wool. Mouton lamb collar. $24.25 [$150-185] **Horsehide in B-15 Bomber** model, lined with sheepskin, plus mouton lamb collar and sleeve lining rayon and wool. $28.75 [$150-185]

Blanket-lined Whipcord for the farmer, sportsman, and outdoor enthusiast. Oxford gray. $5.85 [$55-60] **Super-tough Twisted Twill** fabric stands up to hard work wear. Gray, bark tan. $7.30 [$50-55] **Corduroy Husky.** Sports-style tailoring for trim, athletic appearance. Chestnut brown, navy blue. $10.95 [$40-45] **Cotton Moleskin Air Force** model made for maximum comfort. $9.95 [$60-70]

Two-collar Motorcycle Jacket. Lambskin collar snaps on, self collar when detached. Lined with heavy wool buffalo plaid. Leather belt, metal buckle. Military shoulder straps. $33.95 [$175-225]

Rayon Check Coating with cord tie belt. Capelet shoulder yoke in back. Pink and gray, gold and gray, or blue and navy blue. $17.98 [$60-65] **Rayon Check Coating.** Two-way classic with jockey cap. Wear with or without self belt. $14.98 [$60-65] **Rayon Faille Fitted Coat** with elastic shirred waist. Cuffs, big pockets. Complete with beret. Navy blue, bright red, medium gray. $14.98 [$65-70] **Water-repellent** rayon gabardine with gay multicolor rayon taffeta check lining and trim. Convertible cuffs. Storm tab. Navy blue, medium gray, beige, bright red. $12.98 [$65-70]

Wool Topper with long-wearing qualities. Rayon twill lined. Navy blue, light gray, light apricot. $15.98 [$55-65] **Menswear Rayon Flannel.** Straight-up-and-down pocket tabs and deep simulated cuffs. $15.98 [$55-65] **The Most Casual** topper imaginable. Wool suede cloth with rayon taffeta lining. $17.98 [$55-65] **Tip-Topper** is approximately finger-tip length. Curved pockets and shawl collar. Wool suede cloth. $19.98 [$55-65]

Zip-front Surcoat. Two-tone effect rayon sheen gabardine, pleats from yoke to bottom, concealed slash pockets, patented adjustable slide fastener at sides. $10.95 [$60-70] **Tropical Jacket.** Rayon in ventilated tropical weave. Pointed yoke back. Two-button side vents, ocean pearl buttons. $7.95 [$40-45]

All-weather Gabardine. Outer shell of tightly woven combed cotton, fully interlined with wool quilted under rayon lining. Zipper front and cigarette pocket. Two slash pockets. Adjustable straps at waist sides. $9.45 [$65-85]

Double-shoulder Quilted-lined Jacket. Sheen rayon gabardine lined with high-luster rayon quilted to wool. Double quilted over shoulders. $14.75 [$55-60] **California-style Blouse.** Self collar. Wide shirred elastic waistband. Two-button adjustable cuffs. $9.80 [$55-60]

Note: Poodle cloth coats were very popular throughout the 1950s. These styles are particularly attractive. Swing styles with one button near the collar are most common. Many are lined with a poodle-print taffeta.

Wool Donegal Tweed with wool zip-out lining. Black and white or brown and white. $24.98 [$75-80]
Wool Fleece with wool zip-out lining. Nude, copper, cotillion (bright) blue. $35 [$85-90]

Paris-inspired Poodle Cloth, the shaggy, loop-surfaced fabric: **Brief and Belted.** Double tucks at side, inverted back pleat below waist. Smoke gray, flame red, sun gold, cotillion (bright) blue. $19.98 [$85-95] **Triangle Silhouette** adds black cotton velvet detachable scarf. Convertible cuffs. $29.98 [$95-100] For those whose pride and joy is a trim figure that does justice to a belted-fitted waist and a jaunty flare below. $29.98 [$95-100]

Clockwise from bottom left: **Glamorous 3/4-length Coat** (about 34 inches) with enticingly fitted lines, wool fleece and acetate satin lining. Cotton interlined. $21.98 [$40-45] **Regal Velvet** with silver-colored embroidery on the wing collar. Black, medium red, deep purple, cotillion (bright) blue. $19.98 [$65-70] ***"Yarn-dyed" Grays*** in wool melton cloth. Red cotton velvet on collar and convertible cuffs. Light gray, oxford gray. $17.98 [$55-60] **Tip-topper** (about 34 inches long) of wool fleece. Big pockets; clever detailing at cuffs. $21.98 [$55-60]

Above: **Tattersal Check** in the loveliest color combinations imaginable: black cotton velvet collar with navy and red, green and red, or wine and blue. Matching bonnet. $16.50 [$50-55] **Sheen Gabardine** in fitted classic style with self belt, inverted back pleat. Navy blue, dark green, dark gray. Matching polo hat. $16.50 [$60-65] ***Left:*** **Pinwale Corduroy** plays up the importance of fashion and style in raincoats. Gold, rust, dark green, blue. $16.50 [$60-65]

Figure-enhancing Back Interest centers on the button-trimmed contour belt. Convertible stand-up collar. $29.98 [$95-110]

Black Velvet Trims Collar and convertible cuffs. Six jet black buttons for closing, seven for dress-up. $29.98 [$95-120]

Sheen Gabardine with multicolor check acetate taffeta lining. It has a magnificent full sweep. Navy blue, dark green, wine red. $16.50 [$65-75]

***Above:* Surcoat Style.** Insulated with wool inner lining quilted under rayon. Deep-furred wool pile collar. $10.80 [$65-70] ***Left:* Surcoat** with pile alpaca/wool lining on cotton back, sleeves and 10 in. at bottom lined with plaid rayon to reprocessed wool, mouton-processed dyed lamb fur collar. $15.95 [$65-70]

Bomber-type Jacket lined with satin-finish rayon quilted to 10 oz. reprocessed wool. Water repellent wool pile collar. $9.65 [$55-60] **Luster Twill Blouse.** Continuous front waistband with knit back, lined with rayon quilted to reprocessed wool, wool pile collar. $11.95 [$65-70]

Tuxedo Front and graceful back. Slub-weave rayon and acetate ottoman cloth. Costume jewelry on the collar, rhinestone buttons on wide wing cuffs. Gray, navy blue, pink, medium blue. $22.98 [$60-65] **Novelty Wool Diamond Check.** Easy shoulders and gentle flare. Velvet on collar and e underside of wing cuffs. Navy blue and white, beige and white, two-tone gold, two-tone blue. $24.98 [$65-75] **Nubby Looped Surface** wool poodle cloth. Natural shoulders, soft-rolling front, adjustable sleeves. Light beige, medium pink, light gold, medium blue. $29.98 [$75-85]

Above: **Tuxedo Front Flared Topper** with removable embroidered medallion on collar. Ice blue, coral red, pastel gold. $15.98 [$50-55] **Simple and Enduring Lines.** Big patch pockets. Nude, ice blue, coral red, navy blue. $18.98 [$50-55] **Figure-defining Topper** with back pleats, curving belt, standup, back-notched collar, and jaunty pockets. $19.98 [$50-55] ***Right:*** **Wide-sweep Topper** with notched stand-up collar, V-trimmed pockets and pointed, button-trimmed cuffs. $19.98 [$55-65]

Above: Wool Iridescent Poodle Cloth. Embroidered "jewel" touch is easily detached. Convertible cuffs. Iridescent taffeta lining. Light beige, medium pink, light gold, medium blue. $22.98 [$55-65] **Wool Iridescent Poodle Cloth** in a topper style with slotted seam above inverted pleat in back and across the bust in front. $22.98 [$55-65]
Left: Run-around Stripes in wool in an alternating combination of muted pastels and shadowy brown. Natural tone, pink tone, gold tone, blue tone. $22.98 [$45-50]

Latest California Styles: Two-tone Check Jacket. Rayon check with solid color rayon gabardine front and back yokes. Two-tone maroon, two-tone blue. $6.95 [$65-85] **Sheen Gabardine** with brilliant contrasting rayon lining. Inverted pleats at chest pockets and back lined with contrasting rayon. Dark blue with red, dark green with gold. $8.55 [$65-85]

***Above:* Soft Wool Suede** with velvet shawl collar and trim on turn-back cuffs. $19.98 [$85-95] **Pure Wool Fleece** ripples front and back, all-around yoke, turn back cuffs, cotton velvet lined collar. $24.98 [$85-110] **Softly Shaded Check** in fleecy wool. Double breasted with club collar and shirtwaist cuffs. Collar and lining of cuffs are cotton velvet. Flare back with yoke. $22.98 [$85-110] **Shaggy Wool Fleece** with fashion-wise stitched yoke, shawl collar, drop shoulder, turn-back cuffs, welt pockets. $22.98 [$95-110] ***Right:* Fleecy Wool Pin Check** in muted tones, matching scarf with fringe. Heather tan, heather blue. $21.98 [$65-75] **Lively Wool Check** a teen can't resist. Collar and dressmaker cuffs of velvet. Brown and white, blue and white. $19.98 [$60-65]

Enough Prime Quality Cashmere in the fabric to give it this expensive appearance without making the price prohibitive. Fine wool adds sturdiness. Nude, navy blue. $450 [$95-110]

see page 172

Cotton Velvet sparkles with glittering rhinestone buttons. Empire black, regal purple, royal blue, coronation (medium) red. $18.98 [$55-65] **Peruvian Alpaca Pile** with convertible cuffs. Cotton interlined. Blond beige, navy blue, charcoal gray. $27.50 [$55-60] **Neat Check** in wool highlighted by velvet accents on collar and cuffs. Black and white, red and black, royal blue and black. $18.98 [$55-65]

Wool Fleece. Fitted style with inverted box pleat and belt in back. $19.98 [$45-50]

Wool Fleece featuring the important 1953 vogue for stitching details on the collar, above the bust, and on the convertible cuffs. $19.98 [$55-65]

Simple Classic Coat silhouette with roll collar and full, turn-back cuffs. $29.98 [$85-110] **Stylish and Distinctly Different** triangle-shaped inset, Queen Anne collar, and fly-away cuffs. $29.98 [$85-110] Both in suede check of red and black, dark gray and white, or royal blue and black.

Wool Boucle with tightly spun, thick, curly nubs on its surface. Beige, ruby red, winter pink, smoke blue. $38.50 [$95-120]

Triple-stitch Detail forms yoke on this wool boucle. Cuffs are convertible to your own length. $37.98 [$95-120] **Side-flare Coat,** pure wool boucle. Convertible cuffs. $37.98 [$95-120] Both come in beige, ruby red, winter pink, smoke blue.

Collar-less Coat buttons close to the neck with double rows of stitching paralleling front opening, adjustable cuffs. Wool fleece. $29.98 [$95-115] **Importance of Stitching** is reflected in a novel design between shoulder and collar and on collar and cuffs. Convertible cuffs. Wool fleece. $29.98 [$95-115] Both in nude, copper, flame red, winter rose, and tropic (bright) blue.

Wool Sheen Gabardine with knitted windbreak wristlets, concealed zipper. Dark green, dark gray, maroon wine, black, Air Force blue. $21.98 [$95-120]

Reversible Warm-up Jacket. Both sides in rich harmonizing colors with striped knit trim. Satin with wool lining. $9.75 [$65-70]

Panel Insert Warm-up. Action-knit panel inserts at shoulders, knit collar, sleeves, waistband, and pocket welts. $10.90 [$65-70] **Cowhide Sleeve Warm-up** with leather pocket welt trim. Outer shell of reprocessed wool Melton cloth, lined with acetate satin. $14.75 [$65-70]

Above: Crease-resistant Rayon. *Firm-textured fine worsted suiting. $15.50 [$60-65]* **Rayon Gabardine.** Yoked front, mandarin collar, rayon twill lined. $15.50 [$65-75] **Silk-like Gabardine.** 60/40 rayon/ worsted, rayon twill lined. $19.50 [$75-80] ***Right:*** **Rayon Gabardine.** Appearance of costly worsted, rayon twill lined. $16.50 [$75-80]

Note: Suits and jackets from the early 1950s are becoming increasingly interesting to collectors. Many women wear the jackets with contemporary pants and skirts. There is a definite trend to collect clothes that are appropriate for work, not only party gowns. Collectors evaluate the item simply by saying, "I would never wear that anywhere." Although all the early 1950s suits and jackets are of interest, highest prices are paid for wool gabardine, summer-weight rayon gabardine, exquisite detailing, and two tones.

Rayon Gabardine Suits: Shawl Collar with a neat back peplum, flared skirt. $13.98 [$70-75] **Proportioned for the Shorter Figure.** $10.98 [$70-75] **Top Styling,** good fabric, and tailoring at a low price. $10.98 [$75-85] **Stylish Slimming Suit**. $11.98 [$75-85]

Gabardine at Its Best. Faultlessly tailored of luxurious worsted wool. $28.95 [$110-125]

Rayon Crepe and Taffeta tunic dress with rhinestone circles. $10.75 [$75-85] **Rayon Slipper Satin** dressy type suit with its own rhinestone pins. $9.98 [$75-85] **Finely Ribbed Rayon Crepe** looks and feels like silk. Collar and front bodice beaded with rhinestones and simulated pearls. $13.98 [$70-75]

Three-piece Outfit in smart, long-wearing rayon suiting. Jacket and skirt with tiny woven pin checks; extra skirt of solid darker color suiting. $10.29 [$110-125] **Woven Glen Plaid Menswear Suiting.** $9.98 [$95-110] **Beautiful Menswear Flannel.** Rayon suiting that looks like expensive wool. $12.79 [$75-85]

Classic Suit and Coat Match-up. Each suit or coat sold separately or as match mates. Jacket length about 25 inches. Coat is fully interlined. 100 percent pure wool gabardine. $32.50 [$95-115] Rayon gabardine. $17.98 [$85-90]

Gabardines: Long, Low Lapels allow you to show off your best neck jewelry. $17.98 [$95-110] **Straight in Front**, full and fluent in back. $32.50 [$95-110] **Crease-resistant** rayon gabardine. $17.98 [$110-125]

Rayon Tartan Plaid Jacket with blue gabardine lapels, reversible solid/plaid vest with removable men's type studs, and solid skirt with close vent pleat. $19.98 [$85-90] **Belted and Pocketed** to please the eye, rayon Glen plaid and rayon lining. $17.98 [$80-85] **Hound's-tooth Rayon** looks like fine worsted. $16.98 [$80-85] **Cotton Velvet Trim** on collar and scalloped bust flange, 80/20 rayon/worsted. $18.98 [$95-100]

Rayon Gabardine Suits: Top Styling, Good Fabric, and tailoring. $10.98 [$75-80] **A Nod to Fashion.** $13.98 [$75-80] **Flattering Cutaway Jacket,** semi-swing skirt. $13.98 [$80-85] **Shawl Collar** goes with neat back peplum, flared skirt. $13.98 [$85-90]

Wool Sheen Gabardine Suits: Subtly Concealed Fullness for stately poise. $35 [$95-110] **Curved Shawl Collar** and a new pocket treatment. $35 [$95-110] **Slimming Flange** curves from shoulder to waist; convertible collar is equally smart open or closed. $32.50 [$95-115]

Best-selling Suit Themes in rayon sheen gabardine: **Intricate Trapunto Work** on the collar. $19.50 [$110-115] **Beaded Trim** reflects the trend in suit fashion glamour. $21.50 [$110-115] **Softness and Curves** feminize this suit. $18.50 [$95-110] **Triple Tier of Stand-up Tabs** at hipline.. $18.50 [$95-110]

Thrifty Three-piece Outfit. Rayon lined coat, 50/50 wool/rayon. $8.44 [$50-55]

Handsome Longie Suit of rayon gabardine. $5.75 [$40-45]

Two-piece Casual Suit of 22/78 wool/rayon gabardine. $6.45 [$50-55]

Sizes
10 to 20

Sizes
8 to 18

Rayon Gabardine Coat with tartan plaid yokes, two-tone coat, slacks. $9.83 set [$85-115 set] **Rayon Leisure Coat** with two-tone check panel in front, slacks. $10.83 set [$85-125 set]

Roll Pleats form the sauciest peplum you ever saw. Swing skirt. $10.98 [$70-75] **Triple-Flap Detail** instead of pockets, straight skirt $13.98 [$75-80] **Self-Fabric V-trim** simulates pockets. $13.98 [$75-80] **Tabs at Bust,** button-trimmed flange at hipline. $10.98 [$75-80] **Sleek-fitting Cutaway Front** jacket and four-gore swing skirt welcome the return of the petticoat $10.98 [$65-70]

Glen Plaid Rayon Suiting. Short jacket snugged by harmonizing simulated-leather belt. $16.98 [$65-70] **Rayon Sheen Gabardine** with removable, washable white cotton pique collar and cuffs. $14.98 [$65-75] **Top Fashion Hit** in rayon sheen gabardine flaunts unpressed pleated peplum from figure-defining waist. $13.98 [$75-80]

Spring/Summer 1952

Smart Check Suit with extra solid color rayon gabardine skirt. $15.98 [$75-80] **Check and Solid Partnership,** rayon gabardine. $15.98 [$80-85] **Glen Plaids** are fashion news year 'round. Slanted patch pockets vie with front tucking for style interest. $16.98 [$75-80] **Menswear Flannel** is rayon, but as soft and smooth as expensive wool flannel. Man-tailored style jacket. $21.50 [$75-80]

Fall/Winter 1952

Classic Suit varied slightly to meet today's mood with a curved front yoke, a tab with a gold-color pin inset with simulated rhinestones, slight contour padding to suggest a curved hipline. Pure worsted wool sheen gabardine. $29.98 [$95-120]

Doll Waisted and Whirl Skirted. Lightly padded contour hips, sweep skirt, two simulated rhinestone pins, worsted wool sheen gabardine. $29.98 [$95-120]

Rayon and Acetate Crepe tailored into the new, softer, more feminine classic suit, lined with acetate and rayon iridescent taffeta. $17.98 [$70-75] **Accurate Interpretation** of Italian silk in acetate, orlon, and wool blend with interesting slub effect. Crossed tabs, simulated pearl pin and buttons, lined with acetate and rayon iridescent taffeta. $24.50 [$70-75]

Semi-Classic Suit with novelty tabs, handmade buttonholes, acetate and rayon crepe lining. $29.98 [$95-110] **Flared Peplum Suit** easily adaptable to all shorter figures. Iridescent acetate and rayon taffeta lining, handmade buttonholes. $29.98 [$95-110]

Acetate and Rayon Sharkskin. Looks, feels like expensive worsted wool, cardigan neckline ends in winged lapels. $13.50 [$85-90] **Rayon And Acetate Gabardine** for convertible collar classic. Simple lines that do wonders for the shorter figure. $10.98 [$75-80] **Crepe Suiting** of rayon and acetate. Stiffened hips and butterfly cuffs. $16.98 [$75-80]

Worsted Pure Virgin Wool. $32.50 [$110-125]

Gabardine of Rayon and Acetate. Easy-fitting, slightly flared jacket doubles as a topper with other skirts or dresses. $12.98 [$85-90] **Sharkskin,** a favorite fall suiting, of fine acetate and rayon, with shawl collar and half back belt. $12.98 [$75-80] **Tuxedo Collar Suit** in rayon and acetate gabardine. $10.98 [$65-70]

***Above:* Our Finest Faille.** Acetate and rayon, unlined jacket gored for smooth fit; collar, cuffs, and under flaps on peplum are fine cotton velvet. $9.98 [$70-75]. **Unlined Bolero** is flattering to hips. Slim 5-gore skirt. $6.98 [$80-85] **Sheen Gabardine.** Button-trimmed flaps, slim 4-gore skirt. $8.98 [$95-110]
***Left:* Our Finest Menswear Suiting.** Shorter jacket with stitched shoulder bands, stiffened peplum, full 8-gore skirt. $12.98 [$75-80]

Fall/Winter 1952

Dignified, Graceful Suit has smooth, un-fussed lines enhanced by rhinestone buttons on tabs at the hips. Convertible collar. $29.98 [$95-110]

Spring/Summer 1953

Cardigan Neckline outlined with a generous application of white nylon. Easy to remove, washes in a jiffy. Rhinestones on tabs at left bust. $29.98 [$95-120] **Nipped Waist, Padded Hips,** and whirling skirt with rhinestone ornament-trimmed tabs. Buttons have rhinestone centers. $29.98 [$95-110]

Man-made Miracle Fabrics: Subtle Two-tone Ribbed Surface woven of rayon, acetate, and orlon. Soft, lustrous finish. $16.98 [$65-75] **Silk-like Slub Suiting** looks and feels like Italian silk. Woven of acetate, orlon, and wool. Rhinestones on self-covered buttons. $24.50 [$65-75]

Top row: **Washable Outfits: Gabardine Three-piece Eton** suit. $4.98 [$40-45] **Two-tone Shortie** suit. $.3.29 [$40-45] **Knit Shirt,** twill longies. $3.98 [$20-25] **Gabardine Leisure** outfit. $3.98 [$40-45] **Linen-like Texture** rayon fabric. $5.98 [$30-35]
Center row: **Print Shirt,** poplin shorts. $1.85 [$15-18] **Practical Two-piece** button-on style. $1.98 [$15-18] **Cotton Crepe Shirt,** gabardine shorts. $1.98 [$15-18] **Smart Two-tone** shortie style. $2.59 [$15-18] **Knit Shirt,** gabardine shorts. $1.98 [$15-18]
Bottom row: **Cotton Twill Sailor** suit. $2.94 [$18-25] **Thrifty Cotton Poplin** longie suit. $2.29 [$18-20] **Rayon Print Shirt,** gabardine longies. $2.98 [$15-18] **Knit Shirt,** gabardine longies. $2.79 [$15-18] **Smart Two-piece** outfit. $2.98 [$15-18]

Sheen Gabardine. Richer, more beautiful, better than ever because it adds 15 percent nylon to the rayon and acetate. $16.98 [$70-75] **Ribbed Suit** woven of orlon, rayon, and acetate. $17.98 [$80-85] **Monotone Suiting** of orlon and rayon with the feel of pure wool. Rhinestones and self-fabric looping trim at collar and right pocket. $19.98 [$70-75]

Botany Pure Worsted Boucle with nubby, curly surface. Tasteful compromise between classic and dressy. $32.50 [$80-85] **Pure Wool Flannel.** Man-tailored type suit. $18.50 [$85-90] **New in Fabric,** new in silhouette: boxy jacket suit in color-flecked homespun type fabric. $19.98 [$95-100]

Tailored to Impart feeling of height and slimness. $12.98 [$95-110] **Straight, Loose-fitting Jacket** doubles as a topper. Tuxedo front, bound cuffs. $12.98 [$95-120] **Silhouette that Shapes** the waist, feminizes the hipline, flaunts a circular skirt. $12.98 [$95-110] **Feminine Softness** revealed in the curved collar, trim at hips. $12.98 [$95-110]

Rich Blend of Orlon, rayon, and acetate woven into a ribbed fabric with a two-tone glow. Molded jacket waistline, gracefully arched front panel at the pockets. $19.98 [$75-80] **Semi-classic** with new fashion interest in lapel curves, rhinestone ornament, and hip tabs. $21.98 [$70-75] **One of Our Dressiest** suits featuring long V insets fashioned in a stitched corded herringbone design. $17.98 [$70-75]

Pure Wool Homespun-type suiting with a bright color-flecked nubby surface, wee-waisted jacket, whirl-wide circular skirt. $17.98 [$80-90] **Light Color Flecks and Monotone** team effectively in the new, widely accepted boxy jacket suit. $17.98 [$80-90] **A Novel Idea** for closing a nipped waist: an overlapping tab with ornamental button. The wide collar is sparked with the handkerchief touch of white pique. $17.98 [$85-95]

Man-tailored Type Jacket, an important part of your suit wardrobe, teamed with straight skirt. $10.98 [$70-75] **Club Collar Jacket** and swing skirt. $10.98 [$70-75] **Peter Pan Collar** suit with closely spaced buttons down front, straight skirt. $11.98 [$70-75] **New Long Roll Collar** button-less jacket and straight skirt. $11.98 [$70-75] **Sweeping to New Heights** in popularity: suit with a hand-span waist, stand-out hips, and full, circular skirt. $11.98 [$75-80]

Cinch-waisted Suit with the wide, swishy, circular skirt with stand-out padded hipline, 100 percent worsted wool gabardine. $29.98 [$95-120] **Dressy Touch** imparted via braid trim on novel triangle tabs overlapping conventional lapels and on the cuffs. $29.98 [$95-120]